MW01609788

media|matter

thinking|media

*series editors:*
bernd herzogenrath
patricia pisters

# media|matter

## the materiality of media|matter as medium

edited by
bernd herzogenrath

Bloomsbury Academic
An imprint of Bloomsbury Publishing Inc

B L O O M S B U R Y
NEW YORK · LONDON · OXFORD · NEW DELHI · SYDNEY

**Bloomsbury Academic**

An imprint of Bloomsbury Publishing Inc

| | |
|---|---|
| 1385 Broadway | 50 Bedford Square |
| New York | London |
| NY 10018 | WC1B 3DP |
| USA | UK |

**www.bloomsbury.com**

**BLOOMSBURY and the Diana logo are trademarks
of Bloomsbury Publishing Plc**

First published 2015
Paperback edition first published 2017

© Bernd Herzogenrath and Contributors, 2015

All rights reserved. No part of this publication may be reproduced or transmitted
in any form or by any means, electronic or mechanical, including photocopying,
recording, or any information storage or retrieval system, without prior
permission in writing from the publishers.

No responsibility for loss caused to any individual or organization acting on
or refraining from action as a result of the material in this publication
can be accepted by Bloomsbury or the author.

**Library of Congress Cataloging-in-Publication Data**

Media matter: the materiality of media, matter as medium/edited by
Bernd Herzogenrath.
pages cm—(Thinking media)
Summary: "A materialist attempt to redefine the concept of 'medium'
by expanding it to include the elemental"—Provided by publisher.
Includes bibliographical references and index.
ISBN 978-1-62892-383-4 (hardback)
1. Mass media–Philosophy. I. Herzogenrath, Bernd, 1964- editor.
P90.M3675 2015
302.2301–dc23
2015004741

ISBN: HB: 978-1-6289-2383-4
PB: 978-1-5013-2010-1
ePDF: 978-1-6289-2384-1
ePub: 978-1-6289-2385-8

Series: Thinking|Media

Typeset by Deanta Global Publishing Services, Chennai, India

# Contents

# Acknowledgments

I would like to express my gratitude to Bloomsbury (in particular the wonderful Katie Gallof and Mary Al-Sayed) for giving me and us the opportunity to publish this book, and to all those wonderful people that contributed to this volume—it has been a pleasure!

Special thanks go out to Yasmin Afshar and Sebastian Scherer, for all the work you have put into this!

I would also like to thank the *DFG* (the German Research Foundation) for their generous support of the *media|matter* conference, which was the original seed of this book.

I dedicate this book to Janna and Claudia, and to the memory of Frank.

Two of the essays in this book had a previous life in online journals, in slightly different versions and both are republished with kind permission:

Bernd Herzogenrath's essay appeared in *Cinema: Journal of Philosophy and the Moving Image* 6 (2014)

Milla Tiainen's essay appeared in *Necsus: European Journal of Media Studies* (Autumn 2013)

# Contributors

**Hanjo Berressem** teaches American literature and culture at the University of Cologne. He has published books on Thomas Pynchon (*Pynchon's Poetics: Interfacing Theory and Text.* University of Illinois Press, 1992) and on Witold Gombrowicz (*Lines of Desire: Reading Gombrowicz's Fiction with Lacan.* Northwestern University Press, 1998). He has coedited, *Grenzüberschreibungen: Feminismus und Cultural Studies* (with D. Buchwald und H. Volkening, Aisthesis, 2001), *Chaos-Control/Complexity: Chaos Theory and the Human Sciences* (with D. Buchwald, Special Issue: *American Studies*, 1. 2000), *site-specific: from aachen to zwölfkinder—pynchon\germany* (with L. Haferkamp Special Issue: *Pynchon Notes*, 2008), *Deleuzian Events: Writing\History* (with L. Haferkamp LIT, 2009), and *Between Science and Fiction: The Hollow Earth as Concept and Conceit* (with Michael Bucher and Uwe Schwagmeier, Lit, n-1\work—science—medium, 2012). His articles are situated in the fields of French theory, contemporary American fiction, media studies, the interfaces of art and science, as well as "nature writing" and ecology. He has just completed two "complementary" books, one on Gilles Deleuze and the other on Félix Guattari.

**Benjamin Betka** works on the relationship between films, brains, and existential desperation. After the first contact with communication and information science he studied English, philosophy, and pedagogics in Lower Saxony and Texas. His main field of research is the intersection between media, ontologies, and concepts of suffering in a (post-?) Deleuzean context.

**Eva Ehninger** is Laurenz-Professor for Contemporary Art History at Basel University, Switzerland. She has taught at the Goethe-University, Frankfurt/Main, Hochschule für Künste, Bremen, Bern University and Jawaharlal Nehru University, New Delhi. She's the recipient of fellowships and grants by the German National Academic Foundation (2002–2010),

Fulbright Commission (2003–2004), Getty Research Institute, Los Angeles (2009), Max Weber Stiftung (2013), Swiss National Science Foundation (2015) and Terra Foundation for American Art (2016). She has published the book *Vom Farbfeld zur Land Art. Ortsgebundenheit in der amerikanischen Kunst 1950–70* (Silke Schreiber 2013) and co-edited the volume *In Terms of Painting* (Revolver Publishing 2016). Recent essays include " 'Man is Present'. Barnett Newman's Search for the Experience of the Self" (2012); "What's Happening? Allan Kaprow and Claes Oldenburg Argue about Art And Life" (2014); "Mobile Criticism. Mike Kelleys passiv-aggressive Institutionskritik", *kritische berichte* 2 (2014), 46–57. Her current research is focused on the norms of photographic representation.

**Lorenz Engell** is professor of media philosophy at the Bauhaus-Universität Weimar, Germany, where he was the founding dean of the Faculty of Media from 1996 to 2000 and is now codirector of the Internationals Kolleg für Kulturtechnikforschung und Medienphilosophie (IKKM), a research institute funded by the German Federal Government. His research interests comprehend philosophy of film and of television, media anthropology, philosophy of the comedy and studies on seriality and causality. Recent publications are *"Fernsehtheorie zur Einführung"* (2012), *"Playtime"* (2010), and *"Körper des Denkens"* (2013, co-ed.). Forthcoming is *"Mediale Anthropologie"* (co-ed.) and *"Film Denken. Essays zur Philosophie des Films"* (coauthor). He is also the coeditor of the *"Kursbuch Medienkultur"* (1998), *"Zeitschrift für Medien- und Kulturforschung (ZMK)*," and the *"Film Denken"* book series.

**Bernd Herzogenrath** teaches American literature and culture at Goethe-University, Frankfurt/Main, Germany. He is the author of *An American Body|Politic: A Deleuzian Approach* and editor of two books on Tod Browning, two books on Edgar G. Ulmer, *The Farthest Place: The Music of John Luther Adams* and *Time and History in Deleuze and Serres*. At the moment, he is planning a project, *cinapses: thinking|film* that brings together scholars from film studies, philosophy, and the neurosciences (members include Alva Noë and António Damásio).

**Florian Hoof** is an assistant professor at the department of Theatre, Film and Media Studies at the Goethe-University, Frankfurt/Main. His PhD thesis "Angels of Efficiency. A Media History of Consulting" was funded by the German National Academic Foundation. He was guest professor at Università Cattolica del Sacro Cuore, Milano (2013); scientist in residence at the Swiss Institute of Technology (ETH), Zurich (2010), at Australian School of Business, University of New South Wales (2013) and at The University of Sydney Business School (2015). His research interests include Sociomateriality Studies (www.sociomateriality.de), Science and Technology Studies, Media History, Industrial- and Sport-Films. Recent publications: *Engel der Effizienz. Eine Mediengeschichte der Unternehmensberatung*. Konstanz University Press (2015); "Medien managerialer Entscheidung. Decision-Making 'At a Glance'," *Soziale Systeme* 21,1 (2015); with S.K. Boell, "Using Heider's Epistemology of Thing and Medium for Unpacking the Conception of Documents: Gantt Charts and Boundary Objects," *Proceedings of the 12th Annual Meeting of the Document Academy* 2,1 (2015), 1–14.

**Thomas Köner** studied at Musikhochschule Dortmund and CEM Studio Arnhem. He is a distinctive figure in the fields of contemporary music and multimedia art. For more than three decades his work has been internationally recognized and he has excelled in all the areas of his artistic activity, receiving awards such as Golden Nica Ars Electronica (Linz), Transmediale Award (Berlin), Best Young Artist at ARCO (Madrid), and many more. His familiarity with both the visual and sonic arts resulted in numerous commissions to create music for silent films for the Auditorium du Musée du Louvre, Musée d'Orsay, Centre Pompidou, and others. Likewise, he created installations for diverse situations, for example, International Symposium on Electronic Art (ISEA), the Museo Nacional de Bellas Artes Santiago de Chile, to name a few. His works are part of the collections of significant museums such as Musée national d'art moderne, Centre Pompidou Paris, Musée d'art contemporain, Montréal.

Köner is continuing his close relationship with sound art by creating radiophonic works for the National Radio in Germany (Deutschlandradio Kultur, WDR Studio Akustische Kunst,), while also working as a live performer and composer and producer. His music compositions from the early 1990s, including albums Permafrost, Nunatak, and Teimo, were considered pioneering in the field of minimal electronic and are still in print. Köner's acclaimed production skills with his more beat-oriented duo Porter Ricks, whose album Biokinetics is considered "a classic of techno sound," resulted in remix commissions for a.o. Trent Reznor's Nine Inch Nails.

**Katerina Krtilova** Katerina Krtilova is the coordinator of the Kompetenzzentrum Medienanthropologie (Competence Center for Media Anthropology) and researcher at Bauhaus-Universität Weimar, where she also holds a PhD in Media Studies. In 2013/2014 she initiated and coordinated the DFG funded research project 'Positions and Perspectives of German and Czech Media Philosophy'. Since 2014 she is a member of the editorial board of the Internationales Jahrbuch für Medienphilosophie (International Journal of Media Philosophy). She studied Media Studies, Philosophy and Humanities in Prague and Regensburg and worked as a research assistant at Charles University in Prague. Recent publications: 'Gesten des Denkens. Vilém Flussers Theorie des Gesten als Medienphilosophie', in: T. Hildebrandt, F. Goppelsröder, U.Richtmeyer (ed.), *Bild und Geste. Figurationen des Denkens in Philosophie und Kunst*, Bielefeld 2014; 'Intermediality in Media Philosophy', in: B. Herzogenrath (ed.), *Travels in Intermedia(lity). ReBlurring the Boundaries*, Dartmouth 2012.

**Sebastian Scherer** completed his MA in art history and American studies at the Goethe-University, Frankfurt/Main in 2010. He subsequently taught International Journalism at the University for Applied Sciences in Darmstadt. Since 2011 he works as a scientific assistant for the American Studies department at Goethe-University, Frankfurt/Main, where he teaches classes on American Avant-garde Music, the Electronic Frontier, and American Landscape Painting and Photography. His research interests include sound studies, modern and contemporary art, music, and film, as well as audio engineering, and American countercultures.

Currently, he is working on his PhD project on the artistic strategies of Christian Marclay.

**Walter Seitter** was born in Austria and studied philosophy, political sciences, history of art in Salzburg, Munich and Paris. He translated many philosophical books from French to German. In Aix-la-Chapelle he wrote his habilitation thesis *Menschenfassungen*, and since 1985 he has been teaching media theory in Vienna. He is an active founding member of Vienna's First Philosophical Coffeeshop, the New Vienna Group/Lacan School and the Hermesgroup. His main fields of interest are Political Theory, Philosophical Aesthetics and Physics. Publications include *Poetik lesen* 1, 2 (Merve Verlag 2010, 2014), and *Menschenfassungen. Studien zur Erkenntnispolitikwissenschaft.* (Velbrück 2012).

**Garrett Stewart**, James O. Freedman Professor of Letters at the University of Iowa, has authored, among many books on literature and film, including the *Closed Circuits: Screening Narrative Surveillance* (2015), three studies of reading and its figuration: *Dear Reader: The Conscripted Audience in Nineteenth-Century British Literature* (1996), *The Look of Reading: Book, Painting, Text* (2006), and *Bookwork: Medium to Object to Concept to Art* (2011), all published by the University of Chicago Press. He was elected in 2010 to the American Academy of Arts and Sciences.

**Milla Tiainen** completed her PhD at the University of Turku, Finland, and works currently as a Senior Lecturer in the Department of Musicology at the University of Helsinki. Her research interests range across musical performance, the voice in contemporary arts and media, cultural and feminist musicology, and new materialist and posthumanist thinking. In addition to a monograph and a co-edited collection of essays (in Finnish), she has published widely in international edited volumes and in journals such as *Body&Society*, *Cultural Studies Review*, and *NECSUS – European Journal of Media Studies*. She is completing a book about a new Deleuze-inspired approach to operatic performance (under contract with The University of Minnesota Press). Milla is a founding member of the COST-funded research initiative "New Materialism: Networking European Scholarship on How 'Matter Comes to Matter?'" (2014–2018).

**Stephen Zepke** is an independent researcher living in Vienna. He is the author of *Sublime Art, Towards an Aesthetics of the Future* (EUP 2016 forthcoming) and *Art as Abstract Machine, Ontology and Aesthetics in Deleuze and Guattari* (Routledge 2005). He is the coeditor (with Sjoerd van Tuinen) of *Art History After Deleuze* (Leuven 2016 forthcoming), and (with Simon O'Sullivan) of *Deleuze and Contemporary Art* (EUP 2010) and *Deleuze, Guattari and the Production of the New* (Continuum 2008).

# Media|Matter: An Introduction

Bernd Herzogenrath

The philologies and the humanities are part of the modern media society, and thus participate in a field of forces determined by constantly changing parameters, be it cultural, material, theoretical, scientific, or technical ones. Since the growing proliferation of media and new technologies does not leave the field of the humanities unchanged, they have answered that challenge by reorganizing themselves as cultural studies and media studies. This basically calls for a thorough investigation into and analysis of the concept of "media," which is even more desirable and necessary, since the largely mono-causal attempts at defining the term "medium" have left such an analysis still waiting in limbo.

Media transmit, save, and symbolize. They communicate an "Other" that evades direct access, something that is neither only sign, nor only perception, nor only representation: media are the *in-between*, that which facilitates transmission and thus belongs to neither of the (at least) two sides between which it transmits and negotiates—a status of difference that makes a general definition of "medium" and "media" next to impossible and thus has so far resulted in a myriad of converging attempts to do exactly this. In a formula reminiscent of a notorious Lacanian aphorism, Claus Pias, Joseph Vogl, and Lorenz Engell state in their *Kursbuch Medienkultur* that "media do not exist, at least not media in a substantial and historically stable sense" (Pias et al. 2004: 10). The term "medium" is used to refer a.o. to biological, physical, technological, institutional, and aesthetic media—a theoretical positioning that oscillates between physical properties, qualities, technologies, materialities, and artistic means of expression. The term medium thus encompasses the realms of both culture and nature—and still, current definitions mostly highlight medial aspects of symbolization and information transfer on a cultural level, instrumental and organizational functions, that is.

Hence, the history of media (and of media studies) presents itself as a history of man-made apparatuses and technologies. However—just as the field of cultural studies is currently experiencing a "turn," so also media studies are about to re-adjust, re-structure, re-orientate (and in some circles, e.g., the IKKM Weimar, this new turn has had very promising and fruitful effects already—this volume thus also serves as a part-time introduction to the field of "New German Media Theory," or "German Media Philosophy," of which people such as Lorenz Engell, contributor to this volume, are among its prime agents and innovators). I am not so much referring to a widening of a historical perspective that extends the "medial perspective" from modernity and the digital age also to antiquity and the early middle ages.[1] Such a perspective allows for a clearer distance vision, but such a version of media studies would still be monocular, would still have a blind spot. In contrast to a media studies indebted to the "Holy Writ" of cultural studies (cultural, social, linguistic constructivism, that is), a more inclusive version of media studies would also have to focus on what cultural studies bracket off or only see as a retro-effect of the culturally constituted construction of the world: matter, materiality, or, to call it by its name usually safeguarded by quotation marks—nature. With such terms—matter, materiality, nature—I by no means want to humor any notion of substance. This is not about a return to some form of essentialism, nor about the substitution of discursive "cultural laws" by "natural laws" (be it physical, biological, etc.), but about the modeling of systemic processes and dynamics that underlie both culture and nature: the materiality of biological and social systems seen as self-organizing aggregates that allow for the emergence of newness. For a re-adjustment of media studies, this would have a twofold consequence: on the one hand, to extend the understanding of "medium" in such a way as to include a concept of materiality that also includes "non-human" transmitters (e.g., the elements such as water, earth, fire, air … thus the biological|physical notion of the term "medium"). On the other hand, it would mean to understand media not only in the context of cultural or discursive systems or apparatuses, relays, transistors, hardware, or "Aufschreibesysteme" (discourse networks), but more inclusive in terms of a "media ecology" (McLuhan), so as to proceed, as Régis Debray puts it, "as if mediology could become in relation to semiology what ecology is to the biosphere. Cannot a 'mediasphere' be treated like an ecosystem, formed on the one hand by populations of signs and on the other by a network of vectors and material bases for the signs?" (108) an ecology that acknowledges the interdependency of cultural *and* natural media. Rather than exclusively concentrating on a "technology *a priori*" of a "total media link on a digital base" (Kittler 1999: 2), as proposed by Friedrich Kittler, one should focus on those excluded fields that a Media Theory conceived as a Theory of technology cannot grasp: again, I am quoting Pias, Vogl, and Engell from their logbook on Media Culture: "Media cannot be reduced to forms of representation such as theater and film, nor to technologies such as printing or telecommunications, nor to symbolic systems such as writing, image, or number" (Pias et al. 2004: 10).

Niklas Luhmann's differentiation between "medium" and "form" points the way. By making this differential bear on the conceptual tandem "loose coupling" and "tight coupling," the loose coupling of the medium becomes the "condition of possibility" for the more concrete constellations and actualizations of the "tight coupling" of form—via this "detour" in defining media, Luhmann is able to grasp something of a more "materialist" meaning of the term medium which the exclusively technological and cultural versions of Media Theory have rather neglected. As a sociologist, Luhmann is mainly interested in social systems. In Fritz Heider's book *Ding und Medium* (Thing and Medium), however, from which Luhmann takes over the medium|form distinction, Heider uses the example of sand: sand as infinitely divisible, as a medium that functions "for" a form. The sand's loose coupling, its grain, can serve as a "carrier" for different traces, but can also generate different "forms": sand castles, sand jets, etc. The sand consists of loosely coupled elements (small stones) that are structured by something more solid that has been imprinted on it, the foot. The sand becomes a medium because of the form imprinted on it, and the footprint as a form exists only because of the sand as medium. Through one imprinting foot, the sand elements are momentarily tightly coupled in one way. Through another they become fixed differently. Only because the sand elements are so loosely coupled relative to the imprinting foot (the small stones do not fuse together into a solid mass for instance, making it possible for human feet to alter the coupling of the small stones), they are able to function as an effective medium of indefinitely many discrete footprints.

A "double interface" of this medium|form assemblage makes it quite fruitful for a more inclusive concept of media, I would argue: on the one hand, there is the connection to the radical constructivist theory of autopoietic systems (the theories of Humberto R. Maturana und Francisco J. Varela, to which Luhmann's Systems Theory refers), on the other hand, there is the recursive structure of the medium|form assemblage—in the medium of language (loosely coupled phonemes)— words are the tightly coupled form; in the medium words, sentences are the form; in the medium sentences, thoughts and texts are forms … media thus always are forms of more "loosely coupled" media.

This structure can serve as a system-theoretical variant (and confirmation) of McLuhan's dictum that "the 'content' of any medium is always another medium" (McLuhan 1964: 23). With regard to a wider conception of the term "medium," this means that the medium|form differential spans nature and culture—sand|sandstone— sandstone|sculpture—sculpture|exhibition—exhibition|art scene, etc. … a recursive structure, which is open to both sides and which constitutes mediality always already as (structural) *inter*mediality. Knut Hickethier's (rather dismissive) question in his introduction to media studies— "Should Media Studies deal with the medium 'heat' or 'sand'?" (19) should be answered (with Luhmann and beyond) with a clear "Yes!"

Such an extended concept of media also questions the relation of media and world, or environment—do media "mediate" an access to the world (and thus stand in opposition to the world), or are they a coextensive part of that world? Related to this issue is the question if media are more or less neutral and innocent transmitters of a merely external information, or if media as matter is not itself "informed": the material qualities of a medium, one might argue, are its information, an information that is transferred by coming into touch with another (itself informed) material, and thus both information are changed. They "inform" each other—as Deleuze and Guattari put it: "Matter … is not dead, brute, homogeneous matter, but a matter-movement bearing singularities or haecceities, qualities and even operations" (1987: 512). Such processes are in operation continuously—they are never unidirectional and monocausal, but reciprocal and feedback looped, and they traverse both nature and culture. Media thus are always information *and* matter, that is *Eigen*information (what Georg Simmel has beautifully called the "Eigengesetzlichkeit des Materials"— the "laws governing the material" (1959: 259), its "material qualities"), plus "external information" (the information stored and transmitted by these media)—media as an oscillation and communication between informed matter and materialized information.

Hence, all modes of matter (with all the implications of the medium|form assemblage) can count as media: sound, air, light, animals, human beings, communities, and complex technical machines—Luhmann's structural intermediality might thus be a special case of a more fundamental material intermediality embedded in the interdependency of nature and culture, bio-media, and scriptural|technological media—intermediality as an inclusive media ecology. Such an understanding of media cannot be covered by the humanities single-handedly—a concept of media as matter, as "qualities and agencies" of matter, explicitly calls for a transdisciplinary approach. If Media History in its version as the History of Apparatuses has defined transdisciplinarity simply within the borders and confines of the humanities and cultural studies (with Kittler breaking through into the field of cybernetics and the history and science of technology), leaving fields such as biology, physics, chemistry aside, an expanded concept of media needs to open a dialogue with complexity theory and the theories of nonlinear systems that, based on a shared concept of scientificity, deal with the complex dynamics of both cultural and natural systems and their feedback loops within a *nature|culture-continuum*.

At a time when the difference between nature and culture is being eroded more and more each day, when nature is being "fabricated" and technological innovations are being adapted from nature, at a time when "any distinction between nature and artifice is becoming blurred" (Deleuze 1995: 155), art and culture cannot be described only from the vantage point of a cultural studies sure of itself and its position. As Deleuze and Guattari state:

> We make no distinction between man and nature: the human essence of nature and the natural essence of man become one within nature in the form of production or industry ... man and nature are not like two opposite terms confronting each other—not even in the sense of bipolar opposites within a relationship of causation, ideation, or expression (cause and effect, subject and object, etc.); rather, they are one and the same essential reality, the producer-product. (1992: 4–5)

From this perspective, cultural and media studies are asked to extend their domains, to focus on culture and media in the much more inclusive force field of a "media ecology" that deals with media not only in Kittler's "total media link on a digital basis" and a more general intermediality, but that also allows matter, the materiality of media, its place in that *inter*, in that ecology.

## Theory|Matter

According to Walter Seitter, Aristotle was the first theorist of media. He defined the media as materials "between" the perceived and the perceiver: light and air are natural media for the perception and their analysis is subject of physics. His famous *Metaphysics* extends the field of physics to all material things—also the artificial ones: for instance, the house as a set for putting up human beings, animals, gadgets.

The title "medium" has been given to the house by Marshall McLuhan and the general definition of medium suggested by these two authors could be "a means of presentation" (in difference to a means of production). The mere function of presentation condemns media often to a very weak or inconspicuous existence: they work for the existence of the other. On the other side they have sometimes an excessive impact on their subject: the medium is the message and even the massage. As Seitter explicates in his essay, media are manic-depressive entities. So the physics of media has to consider a wide spectrum of things, sets, processes. Seitter's essay also conjures up the likes of Bruno Latour (with a special kind of key) and Michel Foucault (with "statement" and "discourse") as accomplices.

On the one hand, media come to matter in the historical analysis of cultural techniques. On the other hand, in media philosophy, media come to matter as the mediality of reflection. While media theories like those of Friedrich Kittler, disputing the concept of culture as a sphere of ideas (the human spirit—"Geist"), focus on the materiality of cultural techniques, Katerina Krtilova's media-philosophical approach questions technologies as the condition of possibility of any discourse. In media philosophy, materiality is rather a "force of resistance" (Derrida 2000: 350)—a material resistance in the process of signification and, more radically, completely independent from any symbolic order, preceding anything significant, "something as something" (Mersch 2002: 12–13). Beyond the opposition of a "positive" and a "negative" notion of materiality, the article suggests that media practices be understood using a concept of performativity between those of Judith Butler and Dieter Mersch, proposing a performative concept of reflection.

## Text|Matter

In extending his previous study of the conceptual *bibliobjet* (the "demediated" book sculpture) into a new-media format of post-Gutenberg "print matter"—namely, the actual computer generation of 3D "matter" in so-called digital printing—Garrett Stewart's essay reviews, and extrapolates from, the assumptions of his earlier work on displaced conceptual "legibility." In traditional print culture, what matters in reading the so-called substance of a page, at least in bound form, depends on a substrate of materiality often unrelated (though necessary) to it. An entire genre of conceptual art takes this as its premise and its provocation. One sculptural piece after another, often taking a codex to pieces, offers a thought experiment in the contemplation of this in/dependency of materiality and subject matter. The loss of the latter, via the erosion or occlusion or dysfunctional simulation of text, even when offering a foregrounding of the former surface matter, still leaves the book no real ground to stand on as textual medium. Meaning comes otherwise, troped by deformation. A similar result attends certain recent experiments in 3D printing, whose interest far exceeds the phenomenon of their own technology. Together with the text-suppressed bookwork, these objects explore—by inhabiting—the overlapping space between mediations. Like the *bibliobjet*, freed from the fetish of policed "specificity" in high modernism, such works are nonetheless deeply media specifying in their mixture of modes—and hence operatively "transmedial" in their impurity.

Drawing on theories of light from Fritz Heider to Gilles Deleuze, Hanjo Berressem's essay redefines the genre of "local color writing" as "local light writing" by way of a reading of the inherently luminous poetics of William Faulkner's novel *Light in August*. Local color writers tend to have firsthand experience of living in the geographical regions their stories are set in. In other words, their stories are written from within the field of what Francisco Varela and Humberto Maturana call "structural couplings." Developing their stories directly from the milieu, and thus according to a poetics of immanence rather than to a poetics of representation, local color writers express rather than represent specific milieus. An important element of that expressionism is the light that suffuses these regions and that forms the optical medium in which these regions are given. Drawing on examples from Faulkner, the essay argues that the notion of "local color" is not adequate to conceptualize the complex ways in which writers such as Faulkner use the notion of light. While it is quite natural for the visual arts, and in particular for the "luminists," to use light as their medium of choice, the essay argues that one can also talk of a literary luminism.

## Film|Matter

Following Bill Morrison's photo-essay, Bernd Herzogenrath's essay will focus on Morrison's work and the nexus of film, time, and materiality. The essay begins by introducing film's constitutive|constituting move as the attempt to *represent* time *in* film, which was already being discussed at the birth of the medium. Taking his cue from Bazin's influential article on the "Ontology of the Photographic Image," Herzogenrath shifts his focus to the *materiality* of film: time leaves much more direct traces *on* film than any representation of time *in* film could ever achieve. Taking Bill Morrison's film *Decasia* (2002) as example, Herzogenrath directs a more "materialist" approach to the filmic *material*.

Material culture is based on the premise that the *materiality* of objects is an integrative part and parcel of culture, that the material dimension is as fundamentally important in the understanding of a culture as language or social relations—but material culture mainly focuses on the materiality of everyday objects and their *representation* in the media (literature, film, arts, etc.). Thus, a further and important step would be to redirect such an analysis to the materiality of the media *itself*, to put the probing finger not only at the thing *in* representation, but also at the thing *of* representation. The medium "film" seems to be most promising to test such an interface of material culture and media studies, since film has entertained a most complex relation to *time* from its early beginnings onward: film promised to (re)present temporal dynamics—and the temporality of things—*directly*, un*media*ted, a paradox that gives rise to the different "strategies" of what Deleuze calls the *movement-image* and the *time-image*, respectively. Such a representation, however, is not only an effect of a perceptive illusion, but also of the *repression* of the very materiality of film itself.

If such an interest in the possibilities of the celluloid had already driven much of the 1960s' avant-garde (Brakhage, Jacobs, etc.), *Decasia*, in addition, focuses not only on film's "thingness," but also on its own, particular "temporality." Put together from *found footage* and archive material in various states of "dying," this film reveals the "collaboration" of time and matter as *in itself* "creative," and ultimately produces a category that I will call the *matter-image* and that, Herzogenrath argues, neither Deleuze's *movement-image* nor his *time-image* completely sufficiently grasps: here, time and matter *produce their own filmic image.*

In recent research in philosophy and in the humanities, there is a strong tendency toward questions of matter and materiality, from "Object-oriented Ontology" through "Agential Realism" and "Material Culture." Lorenz Engell's article looks for a link between these approaches on the one hand, and the aspect of media, and the mediatic, on the other. One surprising parallel can be found in Pier Paolo Pasolini's approach to film theory in his essay on "The written language of reality," originally presented in 1966 and sharply rejected by then influential semiological theories. Re-read today, as the article suggests in detail, not only Pasolini evokes the broad materialist tradition of the medium of film and its theory and aesthetics, but he can also add some important aspects to current debates, such as the aspects of time, and affect. But mainly, Engell's contribution is a call for rethinking ontology from the perspective of film, and, in the light of the moving image as a matter of life, for replacing it by the less metaphysical and transcendental conception of an ontography immanent to the life of matter, and of the human.

Over the last decades, the material dimension of the brain has become a central entity in scholarly discourse. Frequently, as Benjamin Betka argues, the ferocious and destructive elements of the brain are not reflected without bias as they are merely dismissed as disorders. A revised *phrenology*, a myopic focus on the brain as a central mystery, is to be countered to ensure a fertile dialogue between the arts and science. Neuronal plasticity becomes a key concept. In order to acknowledge thinking's full capacities and to overcome untenable anthropocentrism it must be applied with care. While considering the state of melancholy as a naked, minoritarian, but mindful effort apart from categories of usefulness it is possible to grasp brains as vexing entities that enable consciousness to flicker through life as a whole. On a posthuman horizon the brain becomes a node in the interplay of *affects* and *objects*, two key terms that are suggested to help in a constructive acknowledgment of melancholy. For Betka, the art of cinema promises to be a valid partner in this enterprise as it deals with such fundamental categories in time and movement.

Florian Hoof's article critically explores recent trends to include theoretical frameworks from science and technology studies to film and media theory. In this context, the concept of *material semiotics*, an approach that enhances the semiotic model from the domain of signs to material objects and technology is introduced to compensate for existing shortcomings in film and media theory. Material semiotics is perceived as a feasible option to overcome the existing dichotomy in film and media theory, which is characterized through approaches that are based either on the concept of representation or centered on a strong notion of technology. Although, at first glance, these theory imports solve the issue of technological and image-as-text centrism, they also introduce the problem of relational determinism into film and media theory. Drawing on existing criticism concerning the relational determinism of material semiotics as a starting point, Hoof proposes to theorize media from the perspective of the *social worlds framework*. He argues for the concept of a historical media epistemology based on the *boundary objects approach* developed by Star and Griesemer, and then extends the idea of boundary objects to the concept of *media boundary objects* as a means of conducting historical analysis on media. The media boundary objects concept Hoof proposes here is a basic theoretical framework for historical as well as systematical research in media and culture. From this perspective, film and media are not limited to provide for the possibilities for communication or to increase the connectivity in a social system. Instead, they can also be conceptualized as structures that stabilize differences and nonsignificant boundaries in society between entities that do not communicate.

## Art|Matter

Eva Ehninger's essay reconsiders two prominent positions of American body and video art, Bruce Nauman and Vito Acconci. Both artists have during the late 1960s and early 1970s worked simultaneously in the media of performance and video and have produced works that prominently feature their respective bodies as they are mediated through the TV monitor. Their artistic practice allows me to focus on the mediality of corporeal matter itself, as body becomes medium and medium—video—becomes visible in its materiality. During this process both lose their original form, but in turn gain semantic possibilities— their interpretation becomes dependent on the individual viewer, who acts as a corporeal counterpart.

Ehninger's description of the human body as a "cool medium" (McLuhan) also serves as a counterargument to the prevalent art-historical short circuiting of artistic works that can be described as "formless"—meaning that they have lost their material and formal integrity and are in the process of dissolving—with the characteristics of the "abject." This argument, which is presented prominently for artistic positions that stage the human body's material dissolution and disintegration, leads to a semantically charged interpretation of formless art: it becomes a mere illustration of the repressed, marginalized, and ultimately the female. In her essay, Ehninger proposes instead that since the body as a medium is inscribed in the material of the video, it gains opportunities of constant semantic renegotiation that go far beyond a gender-specific reading of its form, or formless attributes.

Through the lens of Deleuze and Guattari's materialist reading of Marcel Duchamp's readymade, Stephen Zepke examines the debate between the philosophical approaches of rationalist accelerationism and material vitalism about the relationship of matter and thought. Contemporary art has evolved to a "postconceptual," dematerialized state that favors signification and representation instead of aesthetic experience and sensation. As a force of intervention Deleuze and Guattari maintain a romantic version of material vitalism evoked by nonrepresentational art, which moves toward an ontology of becoming. This creative act of appropriation inherent in the readymade is illustrated by Deleuze and Guattari's motto, "Take anything and make it a matter of expression" (1987: 349). Duchamp's conception of an object as readymade through nomination by the artist and completion by a spectator is contrasted with Deleuze and Guattari's autopoeitic notion of the readymade refrain, which yields expressive and transformative deterritorializations that eventually reconcile life and art. Zepke renegotiates this rupture between discursive Rationalism and romantic materialism by analyzing Guattari's understanding of Bakhtin's concept of the "aesthetic object" and Hjelmslev's material linguistics. Taking Duchamp's *Bottle Rack* as an example, Zepke claims that representational schemata are transcended as form no longer expresses content but is dynamically constructed by it. Aesthetic creation is thus not to be perceived as a matter of information or communication, but rather as a liberatory politics of minor art that is capable of producing "non-sense" outside our everyday affections and perception and that defies the representational logics of capitalism and accelerationism.

# Sound|Matter

The voice has attracted repeated investigation in theories and differently focused analyses of media across areas from film and digital culture to musicology and philosophy. Projects within and beyond media studies have also speculated about the voice itself as a medium or as involved in crucial processes of mediation.

Milla Tiainen's article aims to reappraise and expand existing theories of voice, media, and mediation by engaging three themes that inform many previous investigations of voice and media commingling. She claims that a further thinking of these themes advances our grasp of how voices figure with media and how they might contribute to understanding media; also, how recent reconceptualizations of media and mediation help to reassess the ontology and activities of voice. The first theme concerns the relationality of voice, while the second has to do with voice as a sensory, perceptual event. The third theme addresses the ontological elasticity of voices that refers to their ever-evolving and reinvented actualizations in mediatized milieus.

Tiainen proposes inflections to these themes with new materialist lines of thought. With regard to relationality, she evokes the new materialist emphasis on constitutive relations and emergence. With regard to sensory experience, she turns to the conceptual pair actual/virtual, which new materialist and resonating projects in media and cultural theory have recalled to revisit the sensing body and the relations underlying perceptual experience. As regards the overall ontological contingency of voice, Tiainen extends the debate by elaborating the notion of vocalizing the posthuman. To demonstrate how theoretical reformulations are generated by situational detail and the singularities of each enquiry, the article develops its arguments with two media cultural examples: the feature film *The King's Speech* (2010) and the comic strip–inspired vocalizations of Cathy Berberian.

Sebastian Scherer's essay examines Christian Marclay's hybrid position between sculpture and music, between avant-garde and popular, material culture. In his unusual approaches to music, performance, and visual arts, he not only refers to the lineage of musical practice and sonic philosophy of Charles Ives, John Cage, and Musique concrète, but he also adapts artistic positions of Marcel Duchamp and the Fluxus movement. As a pioneer of

contemporary musical strategies, like sampling and remixing, it
is thus the vinyl record itself, which Marclay takes to its physical
extremes: he deconstructs and recontextualizes musical readymades,
implicitly fraught with the shared cultural baggage of more than a
hundred years of sound recording history, to provoke rhizomatic
juxtapositions and aesthetic feedback loops which oscillate between
the visual and the acoustic plane. His unpretentious mode of
working may also be rooted in pop-art and punk-culture, yet his
strategies of appropriation go beyond the conventional discourses of
authorship, referentiality, and the crisis of the original. By welcoming
the unwanted artifacts of reproduction and playback, it is rather the
ephemeral, indeterminate processuality in Marclay's works that offers
a fertile soil for new-media art and philosophy alike. His emphasis
on unintentional, nonlinear sounds discloses yet another layer of
experience that grants more immediate access to the materiality
of the appropriated media. By transcending static traditions of
representation and interpretation, Marclay's art not only bears the
potential to challenge prevalent notions of mediality and sound, but
also demands for novel approaches of thinking not only *about*, but
rather *with*, media and sound.

In his short contribution, artist Thomas Köner talks about his use
of media matter, presenting four examples from different areas of his
work—audiovisual installation, performance, video, and a musical
composition. Köner explains why choice does not help and why
boredom for him is a good thing. He talks about unexpected things like
dandruff, and why that helps us understand the obedient numbness of
our admiration of multimedia. We learn that our tools are no means to
escape but part of the prison facilities. At the end somebody dies, and
Köner challenges the idea of continuity.

# Note

1   See the research done, for example, by Kittler, or Faulstich.

# References

Debray, Régis (1996), "Toward an Ecology of Cultures," *Media Manifestos*, 108–32, London: Verso.

Deleuze, Gilles (1995), *Negotiations 1972-1990*, trans. M. Joughin, New York: Columbia University Press.

Deleuze, Gilles and Félix Guattari (1987), *Thousand Plateaus. Capitalism and Schizophrenia*, trans. Brian Massumi, Minneapolis: University of Minnesota Press.

Deleuze, Gilles and Félix Guattari (1992), *Anti-Oedipus. Capitalism and Schizophrenia*, trans. Robert Hurley, Mark Seem and Helen R. Lane, Minneapolis: University of Minnesota Press.

Derrida, Jacques (2000), "Typewriter Ribbon: Limited Ink (2)," in T. Cohen, B. Cohen, J. Hillis Miller and A. Warminski (eds), *Material Events. Paul de Man and the Afterlife of Theory*, trans. Peggy Kamuf, 277–360, Minneapolis: University of Minnesota Press.

Faulstich, Werner (2006), *Mediengeschichte 1. Von den Anfängen bis 1700*, Stuttgart: UTB.

Hickethier, Knut (2003), *Einführung in die Medienwissenschaft*, Stuttgart: Metzler.

Kittler, Friedrich (1999), *Grammophone, Film, Typewriter*, trans., with an introduction, Geoffrey Winthrop-Young and Michael Wutz, Stanford: Stanford University Press.

Luhmann, Niklas (1995), *Die Kunst der Gesellschaft*, Frankfurt/Main: Suhrkamp.

McLuhan, Marshall (1964), *Understanding Media: The Extensions of Man*, New York: McGraw Hill.

Mersch, Dieter (2002), *Was sich zeigt: Materialität, Präsenz, Ereignis*, München: Fink.

Pias, C., J. Vogl, L. Engell, O. Fahle, and B. Neitzel, eds (2004), *Kursbuch Medienkultur. Die maßgeblichen Theorien von Brecht bis Baudrillard*, München: Deutsche Verlagsanstalt (DVA).

Simmel, Georg (1959), "The Ruin," in Kurt H. Wolff (ed.), *Georg Simmel, 1858-1918. A Collection of Essays, with Translations and a Bibliography*, 259–66, Columbus: The Ohio State University Press.

Part One

# [Theory|Matter]

# 1

# [Meta]Physics of Media

Walter Seitter

"Physics of Media" should be read as classification and analysis of media as material entities (Seitter 2002). But does this use of the word "physics" actually correspond to the common usage? Usually, "physics" is about natural or cosmological entities such as atoms, light, magnetism, and the like. The term "media" generally refers to artificial things such as newspapers or TV. Can we call the theoretical study of such things "physics"?

The answer is yes, since even in academic physics there are subdisciplines that examine artificial products like industrial materials, meta-materials, and so on. And Marshall McLuhan has shown that the differences between certain media, like inscriptions on stone or printed books, are material differences.

If we look at the ancient origins of the aforementioned discipline, tracing it back to Aristotle, for instance, we see that he himself called his book *Physics* "Lectures on Nature." Opening the book we see that he does not discuss in detail real phenomena of nature such as mountains, animals, or stars but focuses rather on abstract notions such as form, material, the different kinds of cause, change, movement, emptiness, fullness, time, place. These notions refer to elements, aspects, "accidents" involving bodies. "Body" is the term for that which is substantial in nature (Aristotle 2001: 204b 5), and Aristotle mentions as examples certain "bodies" such as earth, fire, water, air, animal, human being; very rarely does he cite things like bed or wall—that is to say, artificial things, without an explication of their occurrence. Sometimes he indicates that perceptible things are the objects of his treatise (Aristotle 2001: 204b 1, 205a 10ff, 208b 25).

Aristotle is more explicit about the science of physics in those lectures, which, after his death, became known as "Metaphysics"—as if the prefix "meta" had the modern meaning of "meta" in metalanguage: speech about something from a higher level. He declares that physics has the task of investigating and formally defining "sensible" things, in modern English: perceptible things (Aristotle 2003: VII 1037a 12). And he gives more examples of the objects of the physics such as the eclipse of the moon, sleep, calm, threshold, book, breakfast, supper, house. The house is often mentioned in different parts of his physical investigation and so in sum just the so-called *Metaphysics* offers us a scattered but complete physics of the house.

It is the famous book *Metaphysics* that gives up the privilege of natural things regarding the question: what is the range of objects for the physics? Here this field is determined by an epistemological criterion— and the decisive criterion is the perceptibility, which coincides with materiality.

Perceptible or material things can be produced by nature (*physis*) or by art and technique (*techne*). Aristotle also addresses this distinction in his *Metaphysics* (Aristotle 2003: XII 1070a 3ff); but elsewhere he ascribes only the things that have the principle of the movement in themselves to the science of physics (Aristotle 2003: XI 1064a 12ff). So he vacillates between two determinations of the field of physics: only natural bodies or all material bodies.

And the media? Are they exclusively artificial things as we like to assume? Aristotle tends toward the other extreme. His invention of the notion "medium"—articulated as a composite of the article of the third gender (Greek "to," German "das") and the preposition "between": *to metaxy*, hence "the between"—is applied only to the *natural* media of perception; this includes some solid materials, water, air, and light (as an effect of fire or the celestial ether)—and this even though these media are natural bodies and as such more or less identical with the so-called elements. Aristotle does not rule out variations: for instance, if fog or smoke diffuse in between, that is, if a supplementary "medium" enters a certain constellation, then the effect is changed. Painters can imitate such variations—as a transition from natural to technical optical media (Aristoteles 1997).

Of course, Aristotle cannot get around addressing other media, above all media of human interaction, such as language, which is an inevitable medium that can be manipulated. The study of language manipulations is called "rhetoric" or "poetics" or even "dialectics": the art of the more or less correct argumentation. In his book *Poetics*, there are two chapters devoted to the language used to trace it back to its most material elements: the vowel and consonant sounds, the different kinds of words as a "material cause" of the tragedy (Aristoteles 1993: 1456b 20ff). It is true that Aristotle does not call the language a medium—in any case I have not found a statement formally going in this direction. The same is true for the house, but as far as I know there are only two theorists of media who declare the house to be a medium: Marshall McLuhan and Walter Seitter (McLuhan 1996: 126ff; Seitter 2002: 145ff, 2011: 34ff). For McLuhan, the house is a medium as a result of its organo-functional role: with clothing and housing the human being extends his skin, that is, his heat-control mechanism (1996: 119ff). And I see the house as a medium because it stabilizes, modifies our presence: a certain desired presence of human individuals, of the owners or tenants of the house.

Media are conditions, regulations, or even obstacles and in this sense "causes" of presence of something or somebody to somebody. *Media are means of presentation* (and as such different from means of production—or from other things). They can be natural or artificial or even personal: the professor has to be a medium of something for somebody, and so does the actor or the priest.

It is somewhat ironical that in his *Metaphysics* Aristotle enlarges the field of the physics to include low, humble, banal things. And he develops, as it were, the physics of the house in his footnotes: a complex entity, an artifact whose material—wood and so on—refers to nature as origin, but also to nature as its teleology—because the house is a container for gadgets and bodies and these bodies—human or animal—belong to nature; the complexity of the house is such that it is also the main object and the name-giver of yet another science, a so-called practical science: namely, "Economics"—the science of household. So Aristotle enlarges his physics to make it the "physics of all things." Physics of all things and not only of natural things; the physics of the tragedy, for instance, as already mentioned.

Let us consider some of the examples Aristotle mentions as objects of the physics: calm, the threshold, breakfast.

First, the calm—as a quality of wind, which is a quality or a state of the air. The calm is that quality of wind that consists of the absence of wind. What for most of us an irrelevant or pleasant state is for the seafarer an unpleasant or even dangerous one. For meteorologists, it is a symptom of stability and so on. These are some possible qualities and effects of the calm. Its causes are nowadays viewed differently from those in past times. Why is the calm mentioned as object of physics in the *Metaphysics*? Because it is a very good illustration of "the physics of all things"—of all things and even of the tiniest, the most inconspicuous things.

German philosopher Jens Soentgen addressed the subject of inconspicuousness in the nineties of the last century in two books: *Splitter und Scherben. Essays zur Phänomenologie des Unscheinbaren* (1990) and *Das Unscheinbare. Phänomenologische Beschreibung von Stoffen, Dingen und fraktalen Gebilden* (1997).

These statements should not leave one with the impression that a broadened understanding of physics gives way to a completely vague or empty concept of medium. It is the function of presentation that constitutes media as media.

But the inconspicuous has a special relation to the media. More precisely, to the one particular aspect of the media, namely, their peculiar modesty or their "depressive" character—emphasized by Lorenz Engell and Joseph Vogl as follows: "Media make readable, audible, visible, perceptible—they do so, however, with the tendency to eliminate themselves and their decisive participation in those sensualities, and therefore tend to become imperceptible, *anaesthetic"* (1999: 10). Martin Heidegger spoke in 1973 about the "Phänomenologie des Unscheinbaren" precisely in this medial sense (but without explicit reference to the media): "The inconspicuous is what allows what appears to appear, but doesn't appear itself"[1] (1977: 137).

It is true that there is also an aspect of immodesty in the media—emphasized dramatically by McLuhan who spoke of the insubordinate, of the excessive or manic impact of media. Impact, which can be underhand or sensational: "The medium is the message—and even the massage." Media are "manic-depressive" entities.

But now for another Aristotelian example of an object of physics as a medium (in our sense): the threshold as a submedium for the medium house. In antiquity, the threshold was something more important than it is today. It was the holy sphere of limit; it was the very beginning of the house—whose center was the hearth. The modern functional equivalent of the threshold is the key—something that is much tinier, but still important in a similar way. The key is the point of decision, of selection: Who can enter and who cannot? A little gadget that determines access or not-access. The key is a submedium to the medium house and in a certain sense a more typical medium: a miniature that is becoming more and more electrified, electronified, digitalized. The key has certainly been intelligent since the very beginning, because it knows *a priori* its right place, it knows how to fit where and how to function as it should.

Perhaps you are familiar with a special kind of a traditional key invented or used in Berlin, the so-called "Berliner Schlüssel" whom the French sociologist Bruno Latour just dedicated a book to (Latour 1996). This key—connected and disconnectable—compels its user not to forget to lock the door and not to forget the key itself. This key, therefore, has not only a technical intelligence but also a social and, of course, an economic intelligence.

The next Aristotelian example of an object of physics is breakfast. This is certainly an example of a really everyday object, but its categorial status is that of an event. In modern language the two main categories are "thing" and "event." The aforementioned calm also is an event or situation. The breakfast is a cultural event: an event within the framework of an institution. Breakfast is a rather weak institution, ignored by some people, but imposed by other people, for instance, by mothers. Institutions formally have to do with media because they work with language. Mothers impose the daily breakfast on their children by emphasizing how necessary it is for their health. At the supermarket there are many foods such as milk, yoghurt, or muesli with inscriptions referring to the breakfast. We find these inscriptions on the packaging. The packaging or wrapping that encloses modern commodities increases the materiality of the foods. A foodstuff like milk at the supermarket has a supplementary materiality. Its liquidity is protected and hidden and supplemented by the solidity of glass or cardboard, which, in turn, is filled with texts and/or pictures, some informative, others extolling. Milk appears in the form of pieces—the institutionalized event "breakfast" can only occur piecemeal in the form of mediatized things.[2] The modern fate of milk resembles the armor- and shield-fate of the medieval warrior.

Let us look again at the precarious state of things being ascribed to media. A more subtle analysis can be found in a book that ignores the term "medium." In *Archéologie du savoir* Michel Foucault uses the two terms "discourse" and "statement." They allude to a presentation of something by linguistic, graphic, or even pictorial means—let us say: speech, writing, drawing, picture.[3] "Discourse" signifies a greater unity of linguistic or visual presentation, a certain type or direction of such a presentation; "statement" is the smallest unity, the smallest functioning unity (and not some not-signifying phoneme). It is interesting to note that the two English words introduced here etymologically do not refer to linguistic or other representative performances. They refer rather to kinetic or mechanical actions: discourse meaning running back and forth and statement meaning stopping or fixating. Of course, these connotations of the English language cannot be ascribed to the author; but I think that they correspond to what Foucault wanted to say.

With "discourse-statement" Foucault wanted to let appear the pure or the minimal "there is" of "speech"; he wants, in his own words, to "let appear the fact, that there is something ... like speech," or point to "the dimension, that gives or furnishes the speech," "the instance of its appearance," or "its existence itself" (1969: 145ff). And the minimalism of the mere occurrence is emphasized by the dichotomy "appearance-disappearance"[4] (1969: 223).

But why this minimalism? Because a well-functioning medium must be transparent in order to make the object visible, noticeable. And transparency makes the transparent itself rather invisible, this being one of the less well-known principles of optics.[5] To quote Foucault: "The statement neither is visible nor is hidden"; it has that "near transparence," "that quasi-invisibility of the 'there is' ... one needs a kind of conversion of the gaze and of the attitude for recognizing and facing it as it is" (1969: 145).

While he was writing his *Archéologie du savoir*, Michel Foucault stated in an interview:

> What I want to do is simply to make appear what is immediately present and at the same time invisible. My project of discourse is a far-sighted project. I would like to make appear what is too near to our gaze, too near to be seen. ... Catch this invisibility, this being invisible because of being too visible, remove the too neighbouring. ... (2011: 60ff)

The visibility of statement emerges with great difficulty from its opposite or its lack. And its materiality is a very flexible one: it is a repeatable materiality, qualified to produce events, to make series, to form sets, to compose larger discursive unities (Foucault 1969: 134). This is the second, the more sophisticated aspect of the (meta)physical enlargement of the "physics of media"—I call it the "ontological" aspect.[6]

The first aspect was a straightforward extension of physics to encompass all empirical kinds of media: natural, technical, even personal.

# Notes

1  "Das Unscheinbare ist das, was dem Erscheinenden erscheinen ermöglicht, aber selbst nicht erscheint," see Martin Heidegger: "Zähringer Seminar" (1973), in: Martin Heidegger: *Vier Seminare* (1977: 137).

2  For the categorical distinction between thing and medium see Walter Seitter: Ding und Medium bei Helmuth Plessner und bei Fritz Heider, in: Interdisziplinäre Anthropology. Jahrbuch 2, (2014). This distinction does not rule out that both can coincide in one and the same entity. For the medial function of packaging see the chapter "Verpackung, Fassung, Container, Milieu, Institution," in Walter Seitter: *Physik der Medien. Materialien, Apparate, Präsentierungen* (2002: 215ff).

3  Especially regarding "statement" I refer the reader to Michel Foucault: *Archéologie du savoir* (Paris 1969) and regarding "discourse" to the little article "Die Wörter und die Bilder," in Michel Foucault, Walter Seitter: *Das Spektrum der Genealogie* (1996).

4  See Michel Foucault: op. cit.: 223. Some years ago Foucault spoke of "the brute fact … there is some order, … the brute being of order, the order in its mere being, … the naked experience of the order and its manners of being." See Michel Foucault: *Les mots et les choses. Une archéologie des sciences humaines* (1966: 12ff).

5  For this I refer the reader to "Grundsätze der Optik," in Walter Seitter: *Physik des Daseins. Bausteine zu einer Philosophie der Erscheinungen* (1997: 62ff).

6  "Ontology" is the formalistic insistence on the immanent modalities, intensities of being, on the different behaviors or fates of being: between thing and event, between existence and nonexistence. Aristotle proposes this in his *Metaphysics* (2003: IV, 1003a 33ff).

# References

Aristoteles (1993), *Poetik*, Stuttgart: Reclam.

Aristoteles (1997), *Kleine naturwissenschaftliche Schriften (Parva naturalia)*, Stuttgart: Reclam.

Aristotle (2001), *Physics*, Cambridge and London, MA: Harvard University Press.

Aristotle (2003), *Metaphysics*, Cambridge and London, MA: Harvard University Press.

Foucault, M. (1966), *Les mots et les choses. Une archéologie des sciences humaines*, Paris: Gallimard.

Foucault, M. (1969), *Archéologie du savoir*, Paris: Gallimard.

Foucault, M. (1996), "Die Wörter und die Bilder," in M. Foucault and W. Seitter (eds), *Das Spektrum der Genealogie*, Bodenheim: Philo.

Foucault, M. (2011), *Le beau danger. Un Entretien de Michel Foucault avec Claude Bonnefoy*, Paris: éditions EHESS.

Heidegger, M. (1977), *Vier Seminare*, Frankfurt: Vittorio Klostermann.

Latour, B. (1996), *Der Berliner Schlüssel*, Berlin: Akademie Verlag.

McLuhan, M. (1996), *Understanding Media. The Extensions of Man*, Cambridge and London, MA: The MIT Press.

Pias, C., J. Vogl, L. Engell, O. Fahle, and B. Neitzel, eds (1999), *Kursbuch Medienkultur. Die maßgeblichen Theorien von Brecht bis Baudrillard*, Stuttgart: Deutsche Verlags Anstalt.

Seitter, W. (1997), "Grundsätze der Optik," in W. Seitter (ed.), *Physik des Daseins. Bausteine zu einer Philosophie der Erscheinungen*, Wien: Sonderzahl Verlag.

Seitter, W. (2002), *Physik der Medien. Materialien, Apparate, Präsentierungen*, Weimar: Verlag und Datenbank für Geisteswissenschaften.

Seitter, W. (2011), "Menschenfassung Container. Zur Architekturgeschichte und Architekturpolitik eines Neuen Mediums," in F. Böckelmann and W. Seitter (eds), *Tumult. Schriften zur Verkehrswissenschaft. 38: Container/ Containment. Die systemischen Grenzen der Globalisierung*, Wetzlar: Büchse der Pandora.

Seitter, W. (2014), "Ding und Medium bei Helmuth Plessner und bei Fritz Heider," in Gerald Hartung and Matthias Herrgen (eds), *Interdisziplinäre Anthropologie*, Jahrbuch 2. Wien, New York: Springer.

Soentgen, J. (1990), *Splitter und Scherben. Essays zur Phänomenologie des Unscheinbaren*, Dietzenbach: Die Graue Edition.

Soentgen, J. (1997), *Das Unscheinbare. Phänomenologische Beschreibung von Stoffen, Dingen und fraktalen Gebilden*, Berlin: Akademie Verlag.

# 2

# Media Matter: Materiality and Performativity in Media Theory

Katerina Krtilova

Media studies[1] is not only a fairly young discipline, but—as Claus Pias observes in the volume *Was waren Medien?* (What were the media?)—in fact an impossible discipline, at least in Germany: media studies addresses new subject matter *and* questions established methodologies (Pias 2011: 16). "Media" are analyzed as a new kind of predominantly technical objects (from telegraphy to computer technologies) and at the same time all kinds of things, new and old techniques, symbolic systems, practices, or sociotechnical dispositives can be understood as "media." The anthology *Was ist ein Medium?* (What is a medium?) presents a list of examples of what has been described as a medium by Marshall McLuhan, Vilém Flusser, Jean Baudrillard, Niklas Luhmann, and others: a chair, a mirror, a football, a waiting room, a street, an elephant, the election system, faith, or love (Münker and Roesler 2008: 11).

The list of objects that have been described as media in introductions and overviews begins to resemble Borges' Chinese encyclopedia, quoted by Michel Foucault in *The Order of Things*: "Animals are divided into (a) belonging to the emperor, (b) embalmed, (c) tamed, (d) suckling pigs, (e) sirens …" (Foucault 1971: XV). This is not only, in our terms, a *wrong* classification, but also *impossible*, disturbing our system of reference, the link between words *and* things: "In the wonderment of this taxonomy, the thing we apprehend in one great leap, the thing that, by means of the fable, is demonstrated as the exotic charm of another system of thought, is the limitation of our own, the stark impossibility of thinking *that*" (Foucault 1971: XV).

The question, "What is a medium?," plays with the exotic charm of a media theory turned into philosophy: it refers both to *things* described as media and to the concept of "the medium"—questioning both at the same time. We can both "invent" new media by introducing a new concept of media (or "the medium") and find media by analyzing historical facts. Can we think that—media floating between concept and thing?

## "Material" reflections

Media studies based on cultural theory is basically concerned with the material world. In scientific or scholarly practice, we always have to take into account technical devices, materials, time and space arrangements, practices and techniques of writing, notation, storage, drafting, etc. The study of media and cultural techniques opens a new field of research; we could say it discovers "new" materials, practices, and techniques and at the same time finds new perspectives on the subject matter of "older" disciplines (literature, history, art history, philosophy, theater and film studies, anthropology). With regard to media, knowledge, science, theory, and thought are not purely mental or "ideal" processes, but are intertwined with material practices, ranging from bodily movements to complex sociotechnical apparatuses. Moreover, from a media-philosophical perspective, media theory and philosophy are themselves intertwined with material practices. However, this approach does not only introduce practices that were not considered before, but also implies a new kind of methodology, working with ideas connected to, resulting from, and involved in certain cultural practices and techniques.

Refering to Jacques Derrida's critique of representation in the context of the history and practice of science, Hans-Jörg Rheinberger has suggested linking media and theory in an "experimental system": "Experimental systems are vehicles for materializing questions. They inextricably cogenerate, so to speak, the phenomena or material entities and the concepts they come to embody" (Rheinberger 1998: 291). The media-philosophical question of a new kind of thinking, not only about, but also *in*, media, pushes the study of material practices of knowledge toward epistemological and metaphysical questions concerning the conditions of possibility of knowledge: Is a reflection intertwined with material and symbolic cultural practices possible?

The starting point is clear: a critique of the objective point of view, which has to be clearly separated from practice—the material world and the actual observer (all subjective "impurities"). Observations are entangled in practices of describing, watching, drawing, organizing, manipulating things, and communicating with people or animals. Referring to the work of Karen Barad (2007: 185), anthropologist Tim Ingold argues that we can observe the world because we are *of* the world. Research thus can be described in terms of practice: "If its method is that of the practitioner, working with materials, its discipline lies in the observational engagement and perceptual acuity that allow the practitioner to follow what is going on, and in turn to respond to it" (Ingold 2013: 4) This approach suggests a turn away from "objective" science that is interested in theoretical models explaining practice—technically applied (by means of not only technical devices, but also social or psychological techniques), these models attempt to govern and *control* practice. What is the aim of "engaged" observation? This necessarily raises philosophical questions—not on a meta-theoretical level (which would separate the observer from actual observation), but inherent in this kind of theoretical practice—When does practice become reflexive? What kinds of practices lead us to observe something, to "reflect" on a very basic level, to compare different possibilities, manipulate things, process, and combine different materials?

Ingold's description of simple and complex practices comes close to a phenomenological approach, an approach that leaves behind the objective point of view. Instead, this approach intentionally takes a step backward, to initial moments, motives, or motions of reflection not already prefigured by certain discourses and objectives. Ingold avoids the trap of Edmund Husserl's "strictly scientific philosophy" (Husserl 1997), which establishes an objective scientific viewpoint to describe something that precedes any objective viewpoint, but fails to address a problem that Husserl's student Martin Heidegger only hinted at (Heidegger 1977: 17): in modern technology, the ideal objective viewpoint is turned into practice—theoretical models that make it possible to manipulate practices are technically implemented in everyday routines. The correspondence between theory and practice is thus outrun by technology. In technoscience, according to Heidegger's critique, theory does not try to explain anything, but to manipulate everything.[2] Heidegger—and Ingold seems to follow this path—rules out the possibility of new technologies becoming part of any practice, not the manipulative practice of technoscience, but part of a *poietic* practice of being-in-the-world, which Heidegger attributes only to art and craft. Yet, just as it is important to reflect on the "objective" logic of technology, this reflection does not have to be situated in a nontechnological world. In fact, we might be able to transgress the recursive operations of technical systems (in cybernetically oriented approaches, this recursivity replaces reflexivity) exactly because thinking in Western culture is entangled in complex abstract practices like writing, which allow not only logical notations, but also deconstructive readings and writings.

Science, too, can involve thinking: "I perceive thinking as a constitutive part of experimental reasoning, conceived as an embodied disclosing activity that transcends its technical conditions and creates an open reading frame for the emergence of unprecedented events" (Rheinberger 1997: 31).

# Media philosophy

Thanks to Friedrich Kittler, Friedrich Nietzsche has become a predecessor of media theory: "Our writing tools are also working on our thoughts" (Kittler 1999: 200). From the viewpoint of media philosophy, Nietzsche's typewriter is not as interesting as his writing reflecting (type)writing—anticipating a "medial" reflection. As early as 1992, Rudolf Fietz used the term "media philosophy" (Medienphilosophie) to characterize Nietzsche's writing (and his concepts of writing, language, and music) in his monograph *Medienphilosophie. Musik, Sprache und Schrift bei Friedrich Nietzsche.* We can learn from Nietzsche, says Fietz, that there is no content independent from media—content always depends on a "materialized form" (Fietz 1992: 11). Nietzsche not only writes about media like music, language, or writing, but also uses words and concepts (metaphors, rhetorical figures) in a decidedly different way; he transforms his thoughts and allows them to be transformed by language, writing, and strategies of musical composition—this is what Fietz characterizes as media philosophy:

"For one thing, Nietzsche never stops commenting (writing) about other matters (say, about other texts). But what's more, and this is his real art, in this same commenting (writing), the status of the 'about' is reflected and as such, sabotages any exclusivity or finality of meaning" (Fietz 1992: 15).[3] Fietz's media philosophy then suggests that we follow Nietzsche's way of thinking/writing. In his approach it is not the typewriter that intervenes in thinking—as a material device that allows more "mechanical" writing, and thus less demanding for hand and eyesight than writing by hand (which was important in Nietzsche's case)—but rather the materiality of the text. This is how Jacques Derrida describes Paul de Man's deconstructive writing:

> The materiality in question—and one must gauge the importance of this irony or paradox—is not a thing; it is not something (sensible or intelligible); it is not even the matter of a body. As it is not something, as it is nothing and yet it works, *cela œuvre*, this nothing therefore operates, it forces, but as a force of resistance. (Derrida 2000: 350)

According to Fietz, a media-philosophical text would in this sense intervene in thinking on several levels, starting with metaphors that pervade "objective" statements, referential language, and the logic of argumentation, which can never be completely separated from metaphorical language, up to and ending with nonphonetic signs like quotation marks or brackets that drop out of the logic of a text representing language. Footnotes, copying texts, transforming, etc., of course, blur the distinction between the materiality of a text and of a typewriter. The medial (in Fietz's sense) or material (for Derrida and de Man) transformation of writing is not just a matter of writing tools and technical devices connected to the alphanumeric code, but also a matter of questioning the very notion of the "material" that is opposed to the "intellectual" or "symbolic": writing oscillates between the intelligible "inside," the interiority (thought or meaning expressed in spoken and written language—presupposing an ideal meaning), and the sensible "outside" (Derrida 1997: 30), the exteriority of actually *performed* acts of writing as well as the practice and *technique* of writing (which requires not only specific skills, but also tools, machines, and materials). Nietzsche's, Derrida's, and de Man's texts destabilize the material/intellectual dichotomy not in a philosophical discussion of concepts and dichotomies (or a theoretical description of writing), but *practically*, unsettling the order of discussing concepts and describing things. They switch from the objectifying mode of reference to the practice of writing, "operations" with grammar, concepts, rhetorical figures, as well as letters and word processors. De Man suggests—and performs—a disjunction between the literal or referential and the figural sense, or, in other words, the "aesthetically responsive" and the "rhetorically aware" reading. The former is the simple mode of saying something, referring to something, the latter comes into play in Derrida's and de Man's texts, using concepts, dichotomies, tropes (especially metaphors and metonymies) so to say against their usual, "transparent" use. The disjunction of both readings "undoes the pseudo-synthesis of inside and outside, time and space, container and content, part and whole, motion and stasis, self and understanding, writer and reader, metaphor and metonymy, that the text has constructed" (de Man 1979: 72).

Derrida stresses: the disjunction is not a literary strategy and cannot be reduced to an "operation of pure knowledge" (Derrida 2000: 351).

"The grammar of the law is a machine of the letter (*gramma*), a letter machine, a writing machine, a typewriter." Derrida quoting de Man: "There can be no use of language which is not, within a certain perspective, thus radically formal, i.e. mechanical, no matter how deeply this aspect may be concealed by aesthetic, formalistic delusions" (Derrida 2000: 352). Derrida calls de Man's deconstruction a "machine-like dis-figuration" (Ibid.: 355) which depends not on a subject or desire, but a material practice, a reflection operating with, in, and against letters, writing, language. "Machine-*like*"—this means grammar is not a machine (nor is de Man): deconstruction cannot be converted into a "media-materialistic" or "techno-determinist" position (of Kittlerian media theories). A deconstructive text does not operate like a computer program. Recursive operations in machines as well as in sociotechnical systems do not destabilize the system; on the contrary, they are necessary for the optimization and adaptation of the system. There is no disjunction in which the materiality of the machine(s) or the code (?) would interfere in the program (if the computer breaks down, so does the program); this is what Kittler points out: in computation the real becomes a manipulable code (Kittler 1989: 71), matter and intellect are united in symbolic-material operations of computing machines.

## Untouchable matter

What matters to media philosophy seems far away from any material practices beyond writing. The disjunction of referential and figural reading and writing which brings forth the materiality of writing Derrida points at (writing about de Man's writing, in the impossible mode of a deconstructive writing about a deconstructive text) would not be possible without the cultural technique of writing, the materiality of writing tools, paper, and other devices, that are themselves not "significant." They are not based on the logic of signification, the specific kind of representation connected to writing, which, for Derrida, dominates Western thinking, a typewriter is not just the "exterior" of an "interior" meaning. "No one can think or interpret the arrangement of letters of a typewriter," as Kittler notes in his later preface to *Discourse Networks, 1800/1900* (Kittler 2012: 120).[4] The materiality of these letters is not a force of resistance—the materiality of machines, material, or working processes does not intervene in computer programs. Following the cybernetic idea, computing makes it possible to abstract from any material conditions or interferences, as Alan Turing points out: "A man provided with pencil, paper and rubber, and subject to strict discipline, is in effect a universal machine" (Turing 2013: 505). Algorithms are what counts—as tools and results, method and object of mathematical-logical thinking, accomplishing a specific abstraction from material practices. The aim is not to understand or just deal with reality, nature, or the human nature, but to find models that make as much reality as possible "calculable," manipulable.

Dieter Mersch argues against this line of media theory connecting materiality and alterity: materiality escapes any kind of logic and thus also the technology of "logocentrism," criticized by Derrida. Universal computing, which for Kittler exceeds the logic of representation (manipulating the real), nevertheless sticks to a logic expressed in Turing's statement: "operating" reality by implementing technical operations (controlling processes, "making things work") becomes possible only if you can *abstract from* the entanglement in practices, relations with others, the embeddedness in the world, etc. This techno-logic requires a detachment from the practice that Tim Ingold focuses on, and from the "life world" of phenomenology. The intertwining with the world precedes any separation of the mind or consciousness from the material world, the body as "Leib" (the lived body, experienced from "inside," not as a thing) (Husserl 1989), the irreducible Otherness of another human being, experiences of the radically other which are powerful without becoming significant. Mersch understands materiality in this sense as a force that makes anything appear at all resisting any subsumption under the logic of signification, any definition by concepts and differences. Any definition of something *as* something ("als was") imposes a certain symbolic order, a discourse, a classification (Mersch 2002: 12–13). Moments of uncertainty, limits of certain symbolic regimes, the interaction with others do not just bring forth certain conditions of knowledge, production, or symbolization—for example, certain tools, aesthetic forms, the letters of the alphabet—but a formless materiality, a materiality without difference, insisting on perception and interaction with other human beings or animals.

Judith Butler opposes this formless materiality to a materiality subjected to "phallogocentrism." With reference to Irigaray, Derrida, and Julia Kristeva, Butler rereads Plato's *chora* as an initial scene of phallogocentrism, the institution of the (in psychoanalytic terms: paternal) symbolic law as a certain kind of thinking, the one philosophy is based on, from which any "bastard thinking"—nonlogical argumentation or an "improper" use of language—has been removed. As the "receptacle," the *chora* is associated with the maternal body, "excessive matter that cannot be contained within the form/matter distinction" (Butler 1993: 38), a "permanent, ... shapeless non-thing which cannot be named" (Butler 1993: 53), an "inscriptional space" (Butler 1993: 52) that itself cannot be described. The symbolic order, as an order of description, is established by separating its fundament: "The receptacle [chora] cannot be exhaustively thematized or figured in Plato's text, precisely because it is that which conditions and escapes every figuration and thematization" (Butler 1993: 41). Butler wants to open up signification to a materiality that is not dependent on it. Mersch's notion of materiality is close to *chora* as the receptacle; the basis; the *arché* of signification; different from signifier and signified; preceding thinking, understanding, interpretation (Mersch 2002: 19). Mersch even speaks of a "receptibility" ("Empfänglichkeit") not already shaped by a certain regime of sensibility (Mersch 2002: 39). As Mersch stresses, materiality cannot and must not be described as something situated *in* the symbolic order, forced into a discourse, logic, or reasonable order of things, forced to mean something. The only way how to approach materiality is in the "negative" way: "What we aspire to find is without symbol or positive statement, and thus there is nothing that could be brought into play, not even as difference, because every statement would already be separated from it and bound to the symbolic" (Mersch 2002: 30).[5]

Materiality cannot be expressed, symbolized—according to Mersch, we can only express that which cannot be expressed or hint at something that which cannot be expressed, or that which withdraws.

Speaking with Butler, we can, however, question this negativity: this way in fact enforces "claims of identity defined through negation" (Butler 1993: 220–1). The symbolic law is instituted by suppressing the receptacle: no improper use of language, no excessive matter beyond language can penetrate the symbolic law, escape logic. Matter can matter only with regard to symbolization—what *can* be said, *can* be signified is clearly separated from what cannot. Negative statements cannot avoid this positive demarcation: "This naming of what cannot be named is itself a penetration into this receptacle which is at once a violent erasure" (Butler 1993: 44). Here Butler turns to a "positive" performativity: "Here it is a question of writing in language of a foreclosure that institutes language itself: How to write in and of it, and how to write in such a way that escapes the full force of foreclosure and what constitutes its displacements can be read in gaps, fissures, and metonymic movements of the text?" (Butler 1993: 197–8).

Referring to Slavoj Žižek's writing: "Is the 'contingency' of language here mastered in and by a textual practice that speaks as the law, whose rhetoricity is domesticated by the declarative mode?" (Butler 1993: 198). Materiality here "emerges within the system as incoherence, disruption, a threat to its own systematicity" (Butler 1993: 39)—a threat, loss, or trauma of the institution of the symbolic law. Mersch pays a lot of attention to paradoxes, distinguishing the logical paradox, which tries to eliminate everything disrupting the logic as nonsense from the catachrestic paradox, which questions the logic itself. He is not transgressing logic, however, but merely delineating the limits of the symbolic order, enacting the impossibility of thinking that. Butler instead stresses the performative dimension of the "iterable practice" (Butler 1993: 220); in a system of reference, identity is constituted by iteration. Žižek himself points out: "… What is overlooked … is that this guaranteeing the identity of an object in all counterfactual situations—through a change of all its descriptive features is the *retroactive effect of naming itself*" (Butler 1993: 216). What is at stake is not the structure of language, but a *practice* that changes in time, connected to political, social, and cultural conditions. It is material beyond oppositions of matter/form and object/referent: "To the extent that a term is performative, it does not merely refer, but acts in some way to constitute that which it enunciates. The 'referent' of a performative is a kind of action which the performative itself calls for and participates in" (Butler 1993: 217).

A deconstructive or psychoanalytic reading (*chora* as the maternal part, suppressed by the paternal institution of a symbolic order) is only one possible way of bringing in different lines of signification, reflexive strategies other than the declarative mode—and thus performing a different kind of thinking. "Diffracting" the declarative or referential mode might be a more precise expression than "reflecting"—referring to Donna Haraway's suggestion. Reflection is associated with a mirror that reflects the light of reason shed by the subject, making things visible (and, in the reflection, the subject itself), whereas "diffraction" brings into play diverted, deflected, bent rays—something that becomes visible as a result of interferences (Haraway 1992: 300). In Butler's terms, penetrating, occupying, and redeploying the paternal language (Butler 1993: 45), contesting the "'possibility of referentiality' … by the catachrestic use of speech that insists on using proper names improperly …" (Butler 1993: 217–18).

## Media practices

Media or rather "the medial" ("das Mediale") can be located in between the material and the symbolic, *physis* and *psyche*, body and mind, thing and concept (Mersch 2010a: 43). Suggesting a "performative concept of reflection" (Mersch 2010b: 206), Mersch does not only refer to the material/symbolic, body/mind dichotomy, but also redeploys it in a performative way: the medial "operates" beyond, or "across," "against the grain of" signifier and signified, concept and thing (Mersch 2010a: 45). The medial thus allows different modes of reflection, not following the symbolic law, based on the distinction of the material and the symbolic. Obviously, (the) medial is not *something*: "There are no media" (Engell and Vogl 1999: 10), as Lorenz Engell and Joseph Vogl state in their foreword to the anthology *Kursbuch Medienkultur*, neither as a thing represented by the term "medium" nor "medium" as mere concept.

The popular notion of media as means of transport, transmission, or translation of meaning or information sticks to the mode of referential language or the logic of representation, characterized by Mersch as the "meta" (or "trans") approach to media—the mode of speaking *about*, in German "*über*" meaning "*above*" (as in "über-tragen"—*metaphorein*): standing above the things or practices described, above the world, not in it. As an opposing term to *meta*, he suggests the *dia* or *per* (like in *performative*) mode: *through, by, in,* and *with*. A "dia" reflection *is* a *medial practice*, but it does not only refer to practices. In this respect it does not make sense to distinguish between medial processes on the one hand and their reflection in other media on the other hand—this would only restore the *meta* position again. *Dia* mediality thus makes it possible to link the study of cultural practices and techniques—as *material* practices connecting the symbolic and the material—with a "medial" reflection that proceeds "across" the opposition of symbolic/material.

Material practices cannot be reduced to a discursive practice—following Mersch's critique of "logocentrism": we must not impose a logic (of representation, but also of difference) on materiality preceding any reflection, any description, or physical manipulation of "materials." Not all practices that can be reflected are reflexive. A "logocentric" viewpoint presupposes a specific reflexivity that disregards many cultural practices—they fall, so to speak, through the sieve of what seems significant for science or cultural theory. The study of cultural techniques, but also of anthropology as outlined by Tim Ingold, brings into focus a whole range of practices involving art and craft as well as new technologies. Thinking must not be separated from these material and symbolic practices and located on the "symbolic" side of the matter/mind dichotomy. Judith Butler's notion of the performative emphasizes the "positive" force of thinking across this opposition: the symbolic law (suppressing formless a-significant materiality) is neither an ideal entity nor a material foundation that determines what can become significant. It relies on actual performances of discursive practices: on actual acts of describing, writing and rewriting, naming, repeating, archiving, visualizing, etc. Reflecting a discourse does not mean to "stand above" and describe how it works—it is always an intervention that *alters* what can be said: a discursive "diffraction." Media begin to matter when we pay attention to *medial practices*— between mind and matter, material and symbolic. The performance of thinking—with Vilém Flusser we might speak of gestures of thinking (Flusser 2014: 24)—codetermines cultural practices that differentiate between the symbolic and the material, mind and body, concept and thing, culture and nature. The symbolic and the material, concept and thing, culture and nature differentiate *in* practices of writing, sketching, ordering, manipulating things, educating people or taming animals, etc. A performative reflection of these practices does look *upon* the material world, but is entangled in the world, and develops along the lines of uncertainty rather than any given meaning and material ground. Media(l) reflection is a theoretical practice in an open field of disjunctions and interventions, instabilities of meaning and material forces.

## Notes

1   In German: "kulturwissenschaftlich orientierte Medienwissenschaft " (media studies based in cultural theory), distinct from the discipline "Kommunikationswissenschaft" (Communication Studies).

2  On a less metaphysical scale, reflexivity has been described as a
   constitutive feature of modern society: the ability to observe practices,
   theorize and adapt practices and techniques (and finally observe the
   observation: reflect and evaluate theories) maintains and sustains social
   systems—described by Niklas Luhmann as *autopoiesis* (Luhmann
   1996). This kind of feedback seems to be outrun in socio-technical
   systems: reflexivity turns into recursive technical operations, which
   work without intentionality, without any consciousness, without
   human beings.
3  Translated from German by Alice R. Christensen.
4  Translated from German by Katerina Krtilova.
5  Translated from German by Alice R. Christensen.

# References

Barad, K. (2007), *Meeting the Universe Half Way*, Durham: Duke University
   Press.
Butler, J. (1993), *Bodies that Matter. On the Discursive Limits of "Sex,"* London:
   Routledge.
de Man, P. (1979), *Allegories of Reading. Figural Language in Rousseau,
   Nietzsche, Rilke, and Proust*, New Haven, CT: Yale University Press.
Derrida, J. (1997), *Of Grammatology*, trans. G. C. Spivak, Baltimore:
   John Hopkins.
Derrida, J. (2000), "Typewriter Ribbon: Limited Ink (2)," in T. Cohen,
   B. Cohen, J. Hillis Miller, and A. Warminski (eds), *Material Events.
   Paul de Man and the Afterlife of Theory*, trans. Peggy Kamuf, 277–360,
   Minneapolis: University of Minnesota Press.
Engell, L. and J. Vogl (1999), "Vorwort," in C. Pias, J. Vogl, L. Engell,
   O. Fahle, and B. Neitzel (eds), *Kursbuch Medienkultur*, 8–11, Stuttgart:
   DVA.
Fietz, R. (1992), *Medienphilosophie: Musik, Sprache und Schrift bei Friedrich
   Nietzsche*, Würzburg: Königshausen & Neumann.
Flusser, V. (2014), *Gestures*, trans. N. A. Roth, 19–25, Minneapolis: University
   of Minnesota Press.
Foucault, M. (1971), *The Order of Things*, trans. A. Sheridan-Smith, New York:
   Random Books.

Haraway, D. (1992), "The Promises of Monsters: A Regenerative Politics for Inappropriate/d Others," in L. Grossberg, C. Nelson and P. A. Treichler (eds), *Cultural Studies*, 295–337, New York: Routledge.

Heidegger, M. (1977), "The Question Concerning Technology," in M. Heidegger (ed.), *The Question Concerning Technology and Other Essays*, trans. W. Lovitt, 3–35, London and New York: Garland Publishing.

Husserl, E. (1989), *Ideas Pertaining to a Pure Phenomenology and to a Phenomenological Philosophy*, Collected Works, Volume III, trans. R. Rojcewicz and A. Schuwer, Dordrecht: Kluwer.

Husserl, E. (1997), "Encyclopedia Britannica Article," in T. Sheehan and R. E. Palmer (eds), *Psychological and Transcendental Psychology and the Confrontation with Heidegger (1927-1931)*, Collected Works, Volume VI, trans. R. E. Palmer, 79–196, The Hague: Nijhoff.

Ingold, T. (2013), *Making: Anthropology, Archeology, Art and Architecture*, London: Routledge.

Kittler, F. (1989), "Fiktion und Simulation," in ARS ELECTRONICA (ed.), *Philosophien der neuen Technologie*, 57–80, Berlin: Merve.

Kittler, F. (1999), *Gramophone, Film, Typewriter*, transl. G. Winthrop-Young and M. Wutz, Stanford: Stanford University Press.

Kittler, F. (2012), "Vorwort zu Aufschreibesysteme 1800/1900," *Zeitschrift für Medienwissenschaft*, 6 (1): 117–26.

Luhmann, N. (1996), *Social Systems*, transl. J. Bednarz Jr. and D. Baecker, Stanford: Stanford University Press.

Mersch, D. (2002), *Was sich zeigt: Materialität, Präsenz, Ereignis*, München: Fink.

Mersch, D. (2010a), "Irrfahrten. Labyrinthe, Netze und die Un-entscheidbarkeit der Welt," in G. Mein and S. Börnchen (eds), *Weltliche Wallfahrten. Auf der Spur des Realen*, 41–56, München: Fink.

Mersch, D. (2010b), "Meta/Dia. Zwei unterschiedliche Zugänge zum Medialen," *Zeitschrift für Medien- und Kulturforschung*, 2: 185–208.

Münker, S. and A. Roesler (2008), "Vorwort," in S. Münker and A. Roesler (eds), *Was ist ein Medium?* 7–12, Frankfurt/Main: Suhrkamp.

Pias, C. (2011), "Was waren Medien-Wissenschaften?," in: C. Pias (ed.), *Was waren Medien?* 7–30, Zürich: Diaphanes.

Rheinberger, H.-J. (1997), *Toward History of Epistemic Things. Synthesizing Proteins in the Test Tube*, Stanford: Stanford University Press.

Rheinberger, H.-J. (1998), "Experimental Systems, Graphematic Spaces," in T. Lenoir (ed.), *Inscribing Science: Scientific Texts and the Materiality of Communication*, 285–303, Stanford: Stanford University.

Turing, A. (2013), "Intelligent Machinery," in S. B. Cooper and J. van Leeuwen (eds), *Alan Turing. His Work and Impact*, 501–6, Waltham and Oxford, Amsterdam: Elsevier.

Part Two

# [Text|Matter]

# 3

# Between Print Matter and Page Matter: The Codex Platform as Medial Support

Garrett Stewart

What matters in reading the so-called substance of a page, at least in bound form, depends on a substrate of materiality often unrelated (though necessary) to it. An entire genre of conceptual art takes this as its premise and its provocation. One sculptural piece after another, often taking a codex to pieces, offers a thought experiment in the contemplation of this in/dependency of materiality and subject matter. The loss of the latter, via the erosion or occlusion or dysfunctional simulation of text, even when offering a foregrounding of the former surface matter, still leaves the book no real ground to stand on as textual medium. Meaning comes otherwise, troped by deformation.

In enunciation may begin a fundamental recognition. Let us start with what is propositionally entitled by this volume (as text rather than book). Book/text: back to that crucial breakpoint in a moment. For now, as soon as one lets the typographic membrane partitioning "media" from "matter" in this collective project grow momentarily porous, one has already begun to singularize the plural—at once by vocal liaison and graphic elision—in the phonic chiasm of a conjoint *mediumatter*. Only fair, since alphabetic language is itself a materialized morphophonemic medium, an audio/visual composite from the ground up.

And of course other media besides language can be just as mixed, as polyvalent in their sensory claims, as is the contingent wordplay—word splay—of even discursive phrasing. On the title page of a volume like this, such *graphonic* conflation is as much a part of the medium as the paper it is printed on and the codex into which it is sewn or glued—or, alternately, as the backlit digital array of its potential e-book or online "incarnation" (if that can any longer be the word for a bodiless stream of alphabetic data). So this is almost a launching allegory, this phonetic merger at the labial *m*. Put it that when the media are separately manifest in their often impure or layered materiality (their *mediumatter*), and isolated in this condition under aesthetic pressure in conceptual gallery objects, it is then that a particular (if not monolithic) sense of medium is sprung from just such skewed conjunctions of function and form. Sprung—but not loose. The very art of such work is often to measure the spectrum of effect within which its given instance falls. Hence, long after the eclipse of medium-specificity in the high modernist or essentialist mold promulgated by Clement Greenberg, we find the continuance in conceptual art (and indeed sometimes the escalated stakes) of comparative *medium-specification*.

And often by way of conflation, what is also found is a dovetailing of means within a given manifestation. A decade back, Ed Ruscha and Raymond Pettibon collaborated on a series of subtitled lithographs called "The End" (2003), in various "states." Two such captioned works converge to orient a whole swath of conceptual book art in this same decade, even though the immediate optical allusion (and illusion) of these pieces is that of adjacent vertical frames, not pages, on an abraded and scratched film strip, worn thin by time and obsolescence. Here is the twentieth century's premiere or at least signature medium—its distinctive imprint medium—caught between serial photograms in what, for a particular but unspecified film, may be the penultimate cell and its closural successor: the lower half of "The End" (in an old-fashioned gothic script) appearing on exit in the vanishing upper frame, with the emergent (but going nowhere) top half of the two monosyllables in the bottom sector of this en route bifurcation. In part, the effect is to record not just the "Fin" of a given (or rather ungiven) screen text but also of film itself as a discretely serialized rather than bitscanned medium, this in view of the new digital hegemony of postfilmic cinema—an end transcribed, as it happens, indeed as it is happening, in yet another, earlier, and now itself gradually eclipsed material form: ink on paper. So that, in this very layering of material self-consciousness, Ruscha's collaboration with Pettibon for these works of worded image is an effort to record the end of medium-specificity itself under a postmodernist departure from purity into what Fredric Jameson calls the "mediatic system" writ large, where aesthetic space in post-1960s' conceptualism is reconfigured as a cross-feed of reflexive medial instances (Jameson 1991: 162).

And "The End" records this finality explicitly, discursively, as well as pictorially. For the two most telling captions from this series of frameline lithographs, these cinegraphs—etched by hand in slanted uppercase lettering, red and black, respectively—read as follows. First: "AS MY EYES CLEARED, I FELT THE DIZZINESS OF NOT KNOWING WHICH END I WAS AT THE END OF." A text (visual or verbal)? An epoch? The uncertainty is secured by the second of such epitomizing captions: "THE ANALOGY BETWEEN THE ART OF THE ARTIST, THE NOVELIST, AND THE PRINTER IS, SO FAR AS I AM ABLE TO SEE, COMPLETE." Especially complete, one might add, if by "printer" we are meant to include the processor of the cinematic print—and if by "so far as I am able to see" the premium on the visible is made clear under idiomatic cover. Which certainly it is.

After "the end" of "the printer," of course, display becomes all digital, all the time. And precisely a decade after these graphic collaborations— as if in a further memorial homage to the art of "the novelist" or "the printer," the bookmaker at large—it is in fact enlargement per se that is the mode of choice in a series of almost trompe l'oeil open books by Ruscha (though too big to fool us about their utility, instilling instead a sense of their surrendered prominence). These are depicted books gargantuan in format and hovering at something like the wall-sized scale of landscape paintings—but with no scenography represented even in the sign language of writing. None but the horizonless plane of marked but uninscribed—which is to say sheerly visual—surface. Never was Michel Butor's dictum more fully realized than in these works, as if they were huge mute footnotes to his claim that the one thing books have inevitably in common with painting is that all books are diptychs (Butor 1996: 55). For this is never truer, or at least never more apparent, than when books have been stripped, demediated, of the very lettering that makes them work.


## Demediation

With these open 2D tomes of Ruscha's, shown first at Gagosian in New York in 2013, their material pages are almost tangible in the weathered and blotched exactitude of their rendering—but bearing no words at all: the works being legible, that is, only as aged codex formats, as if memorialized by the very silencing of any trace of message. Textuality is reduced to matter without signage: a support without a mediation except what it shares—what it *transmediates*—with the planar coverage of paint. As with so much recent work in this vein, elegy for the long reign of the codex is at one with a celebration of its iconic form—and here monumental projection—across the institutions of our culture, complete in this case with marbleized endpapers in the case of *Gilded, Marbled, and Foibled* (2011–12) that further anachronize these tokens of bibliolatry at an almost architectonic scale of simulated stonework. As if in a magnified capstone, one might say, for this recent aesthetic tendency in biblio/graphic depiction, Ruscha's newest work offers pages so large, unwieldy, and weathered, even when not explicitly marmorealized, that they communicate not any particular message but only the magnitude itself of a vanishing cultural hegemony.

Even as these outsize wall works can engulf perception and attention, they do so in their own different key—not like real books in their portability and process. One might well insist that books are visual objects first of all, whether as bound pages or as readable text. First, maybe, but that is not all. Books appeal to multiple senses other than those that their wording serves to name or evoke. The actual hefted codex has a weight and texture, a feel, as well as a look: a look apart from what it pictures. Also, to varying degrees, a smell. Plus a certain resistance, even when there are no longer, in modern publication, any folded pages to be cut (the codex itself now en route to antiquation alongside the bone-bladed page knife). And a certain potential savor as well, apart from text, even for readers not given to licking their index fingers for faster page turning. All of that is gone, to be sure, from the new uses of the digit for electronic reading. And for all the gains in convenience, in "affordance," what the hinged codex used to localize as appendage to the human frame—one articulated material form meeting another over the so-called body of reading—is often what seems delimited and reflected upon now, quite apart from the gradual replacement of its function in commercial publishing (with its vestigial metaphors of "paging" and "scrolling" and "bookmarking"), by the wholesale conceptual defacement of its form in sculptural exhibitions. It is my own book on that latter topic—examining not the traditional artist's book but instead a text-suspended book art descended from the estrangements and *detournements* of conceptualist practice (Stewart 2011)—that I here reopen, along with an early debate about its premises, in order to appreciate a good deal of new evidence coming to my attention, and often into being (like Ruscha's Picture Books), in the half decade since my earlier survey went to press.

Text/book—text versus book: a breakpoint to which I promised a return. And such a return takes me directly back, in effect, to my earlier arguments about what I would now call the textualization of codex sculpture in that wide array of conceptual artifacts that were gathered for inspection under the term *bibliobjets*. Convened there, in *Bookwork*, were some of the many found and defaced, sealed or buried, simulated or mounded volumes produced in recent years so as to be explicitly inaccessible to reading: a reading denied to view by what I set out to trace as the overt "demediation" of their textual delivery system, whether occluded, mutilated, or muted in some other fashion—or refashioning. To put it this way is to recall a former sticking point I continue holding to. Regarding a seed essay (Stewart 2010) that grew into the fuller argument of *Bookwork*, my claims were met—in a solicited response appearing alongside my piece in the same journal (Lurz 2010)—with a decided demurral from a scholar in the field of book studies and media theory rather than conceptualist art practice. John Lurz's terms gravitated understandably, in a conception of bookhood, to the isolated tangibility of the codex form installed by the works I was discussing—books appropriated or troped (the "lifted" or refigured object)—rather than being drawn, as I was, to the further shaped evocations of textual experience itself that such denatured cultural forms tend to engender in the absence of normative linguistic mediation. Hence the misunderstanding. In other words, where I stressed the lost medium of illegible writing contorted by overcompensation into materialist conceits for the site, insight, or weight of reading, Lurz felt I was missing the material experience of the held book in the first place—as its own brand of mediality, of material conveyance. To call these works "demediated," he felt, was to miss something one needs to remember about all books as physical carriers rather than just verbal constructs and referential conduits.

Extending W. J. T. Mitchell's insistence that all media are mixed media (Mitchell 2005) to the codex format as well, Lurz—as partisan of the material turn over against the textual turn—wanted from my essay a more explicit acknowledgment of the book's palpable medium as object of conveyance, in all its hands-on aspects, material and haptic. Yet this was the very acknowledgment that such unreadable bookwork I had been studying (and, in effect, reading) makes inescapable by its very reduction. It is not that it went without saying, but that it was everywhere visible and under analysis in its own particular terms. For there on the gallery plinth or wall or floor, time after time, are positioned the material girth and heft of the codex's medial substrate, no doubt about it, but witnessing in just this way to the marked loss of its primary mediating or transmissive function in the linguistic register. If the book is a mixed medium, physical and graphic as well linguistic, then its radical embodiment as a strictly volumetric object, with all language scraped or vacuumed out, or otherwise closed to view, results from what I would still want to call, pace Lurz, the work of demediation: both in the refusal of wording and in the inhibition of materiality's normal role in the communicative function. Instead of business as usual, the once functional object is, in these works, denuded to fixed material thing and arrested in its informational circuit. By these means, and in the very process of their conversion to a further meaning, the *biobliobjet* puts quotes around "book" so as to induce a veritable thesaurus of conceptual paraphrase for its shape and feel and physical use, the latter including (but of course not exhausted by) its invitation to verbal immersion. These are works of what I would want now to call *codexduality*, where, despite the missing print of the normative page, the remaining compositional format of the work doubles for an implicit text all its own about the deactivated but still imaginatively bracketed reading experience.

## Within the mediatic "network"

In a spirit close to that of Mitchell's insistence on the composite aspects of all mediation, two recent contextualist rather than purist definitions of "medium" converge here, though with no recourse in either case to Jameson's "mediatic system" as a norm of postmodernism all told—and to conceptual art as its litmus test. It is of no little interest for the present concerns that, in the gallery exhibition offering Jameson's chief exhibit in this line, the reflexive ricochet of medial checkpoints in Robert Gober's installation includes, besides earthwork sculpture and a classic landscape painting, also one of the appropriated book objects of Richard Prince, its title uncited by Jameson (165). The effect (also unmentioned as such) seems to me partly synecdochic: serving to textualize the whole discrepant ensemble as a performative commentary on a no longer essentialist material standing-ground for innovative aesthetic conjunctions. Yet long after Jameson, and with no return to his proposals, the two new and all but interchangeable definitions of the medial function that should further orient this discussion do join many recent gallery displays in confirming the broad claim of *Bookwork*. By which I mean its axiom about how the reduction of codex format to sculptural shape (textuality in the everyday sense thus under erasure or in abeyance) is a "demediating" appropriation and alteration—and, at times, in an apparent further sense, with the shredded or crucified book, the book nailed shut or even cremated, a *demeaning* one.

According to conceptual poet and theorist Craig Dworkin, mediums, as opposed to material supports, "consist of the analyses of networked objects in specific social settings" (Dworkin 2013: 32). This is an arresting point. Before eliciting an analytic procedure, mediums constitute one. That puts a distinctly (and aptly) reflexive spin on our sense of their uptake in perception. Similarly, for art critic Hal Foster, "mediums are social conventions-cum-contracts with technical substrates" understood "in a cultural field" (Foster 2013: xi). In each formulation, therefore, we find the same resistance to the cult of high modernist specificity, to strictly material circumscription and autonomy, to any and all predetermined criteria, as well as the same insistence on the cultural or social space of reception. This is where, for Foster, mediums operate "in a differential process of both analogy with other mediums and distinction from them" (xi). More particularly, what I am singling out as the *transmedial* function of certain bookworks (after the initial demediation of their print basis)—as for instance the figural commute between pulped cellulose and mental picture, concrete phenomenon and textual phenomenology—arises when the determining and most immediate "field" (or "analytic" context) of any one manifold effect is in fact the multiple technical substrates it conflates and reworks. Freed from the fetish of policed specificity, these works are nonetheless, as suggested, deeply media-specifying in their mixture of modes—and hence operatively transmedial in their impurity. And here again Dworkin's sense of mediums dependent on interpretation is worth repeating, on which understanding, again, they "consist of the analyses of networked objects in specific social settings" (32). There is, in short, no medium without social mediation.

If the message carried by a given medium is there for analysis, so be it. But Dworkin is concerned mainly, instead, with the absence of message and the pressure placed thereby on a stable definition (and even dimension) of the medium in play. He begins with an extended reading of Jean Cocteau's film *Orphée*, a poet's crossover text in a new cinematic medium in which, as we discover, another kind of poet's text would constitute the cancellation of poetry and cinema altogether. For in Cocteau's plot, Orpheus' rival poet, Cégeste, brandishes a book of poems called *Nudisme* that amounts, for Dworkin, in its function as "punning rebus" (7), to the *mise à nu*, the "laying bare" (7), of blank white rectangles—failed film frames, we might add, as well as failed verse pages. In this jesting turn, the very gesture, *cegeste*, of nakedness offers no medium but itself. Its blankness thus *ce geste* effects a demediation of verse by sheer exposure (like overexposed film) in the nakedness of verso and recto. Which leaves only the poetry of Cocteau's new cinematic venture, colonizing its own white rectangle with human figuration— my metacinematic point, not Dworkin's—as the medium of note and notice. In Dworkin's close attention to the look of these filmed eventless pages, the recognizable avant-garde poetic format of this (albeit blank) codex and its typical binding is part of the historicized social irony of its gesture, fully rooted (Dworkin can show) in the publishing context of the period. How, then, does such a book of (or for) poems, reduced from text to sheer codex template—yet retaining its metatextual ironies about denuded representation—find itself resonantly foregrounded when framed by another medium equally dependent on serial rectangular manifestation? That seems one rhetorical question behind Cocteau's materialist recession from surrealist screen poetry to blank rectangle.

And it is from this narration on screen of an experimental poetry, a poetry emerging in the form of a demediated *bibliobjet,* that Dworkin moves to actual objects like it in contemporary aesthetic circulation. There is Tom Friedman's famous blank page claiming aesthetic attention not for any application of word or paint to its surface but only for the thousand hours the artist has applied himself in staring blankly at it in the parodic indexicality of this "non-retinal" art (26), where the untreated sheet, Dworkin helps us to see, is the antithesis of the self-developing early "photogram" (and his high modernist revival by the likes of Moholy-Nagy and Man-Ray). Here and in his other myriad examples, Dworkin's evidence is relentlessly to the point. In homage to what one might think of as a founding postmodernist gesture of conceptual art in Rauschenberg's *Erased DeKooning Drawing*, there is Nick Thurston's "edition" of Maurice Blanchot's *The Remove of Literature* (39), all words punningly removed. Or *The Body* by Jenny Boully, a format in which the lost "body of the text" can only be inferred from its surviving "apparatus" in exhumed notes. Gloss has become the only mediating format, flanking the material void of primary discourse, so that none of the standard cellarage of footnotes is ever uncorked in the form of any immanent text. And, again in Dworkin's litany of missing surface text, there is the ream of paper bound as a book but left blank by Aram Saroyan, a book merely in potential, "about to be" (15), and unsigned except in so far as the unspoken ream is a "tidy paragram of Aram" (15).

If for Dworkin a medium-deferred but materially palpable book object can operate a phonetic play on its nonauthorial maker, my own more generalizable sense is that the demediated book object often lays claim to the literary valence its effaces or withholds by just such tacit wordplay. It does this, however, by one mode of displacement or another from a figurative conception of the reading experience rather than from the identification of an authorial signature. When Dworkin's argument in *No Medium* is rounded out by an extraordinary discography of silence in phonographic texts ("Further Listening" 141–73), as with the blank disks packaged as *The Wit and Wisdom of Ronald Reagan* (163), one comes ultimately to suspect an unwritten silent inflection for his own title, as if its real force arrives with an invisible question mark. No Medium? Really? Are you sure? Maybe not where you are looking for it, but there nonetheless because you are (are present, are looking)—and have something to contemplate, to speculate on rather than simply to see or hear, or in other words, even with elided words, something to "analyze" in a "network" with other similar manifestations. Or in the long run to *read*. Just as the museum locus can transform a bathroom object, a mere urinal, into a world-historical *objet* (Marcel Duchamp's 1917 *Fountain*), or in other words an industrial fixture into a trademark modernist sculpture, so, too, can a certain pitch of transmedial attention in book work (applying the paradigms of Dworkin and Foster together to the lineage of Duchamp's conceptualist descendants) educe new aesthetic yield from an abused print artifact or its inert duplication. "Networking" such an object with others bearing on its effect under the same aegis of recognition serves to define the aesthetic as a conceptual field. That is after all, and in short, how a painted wooden block, a bifurcated concrete slab, or set of torched paper sheets can become a *bibliojet*, often figuring in conceptual terms what is has faked or disfeatured in material fact.

## Stress-tested substrates

In all the abrading and fraying and waterlogging, the sanding or pulping, to which the appropriated or *detourned* book is submitted in various installations, two recent instances (unnoted in *Bookwork*) sum up a tendency toward troping the codex as an anatomical structure in its own right, distinct as such from the human communication it services. This somatic figuration was previously exploited by British conceptual artist Fiona Banner when she embossed a bound book under the title "Book Anatomy"—all pages blank—with simply the terms *Face, Spine,* and *Back* on their respective surfaces: a further extrusion of the "body of the text" to the exoskeleton of the codex (Stewart 2011: 149). Since then, in 2013, American conceptual artist Paul Forte displayed the new work *Book of Maladies* in a retrospective exhibit at the University of Rhode Island, where the book's own mobile anatomy had succumbed not just to distress but as if to a disfiguring disease. Moreover, within the micro "network" of this show, the work came under contrapuntal "analysis" (recalling Dworkin's definition) from a companion piece constructed (or "destructed") in the same year, less in this case a bioorganic scourge of the book than an ironic reduction of it to the particulate matter of its own signified vistas.

In this latter case, so we read in the catalog, a found and crumbling book—dated not just by title but in every way—had its pages fall so quickly to pieces in the artist's hands that it could only be preserved by ironic deconstruction. What is left of *Picturesque Palestine, Sinai, and Egypt*, an illustrated travel book from the imperial and colonialist 1880s, can in turn only be "read"—its layered sheets disintegrated to a single sample plane—in the form of a new fragmentary collage, the morcelated and reframed *Desert Parcel* (2013), with its fateful erosion from dust to dust: a granulation strictly material reworked as an implicitly representational expanse once again. Also from 2013, then, and clarified in its resurfacing by contrast with these decimated rather than just defaced pages displayed nearby, that appropriated *Book of Maladies*, title unchanged, appears in the form of its bound but also artificially sealed imprint—further violated (after sanding) by the corrosive effect of applied liquid crystals on its open surface. Any medicinal dimension to this paralyzed and psoriatic volume has thus suffered the fate of its own signifiers: a case, so to say, of deformation following function, though gorgeous in the alchemy of its mottling—and potential anthropomorphic molting. In this malady-ridden volume, book conservation and reconditioning are reversed in the etymological *male habitus* (ill-conditioned) substrate of print impress. Hence, once again the transmediation of the illegible page to the figurative bulk or surface of its own evocation. In each work, content is transferred to form so completely that the body of a text is submitted to the likes of a symptomatic bodily ailment and desert vistas (once captured in descriptive text and photographic panoramas) are turned back over to the sifted grit of their own decay and reconfiguration. Each alteration, verging on disintegration, offers a remade semantics in excess of its former cultural vehicle in codex presentation.

And since then, in a vast multimedia—and, one should add, transmedial—installation at the Haus der Kunst in Munich, American artist Matthew Barney accompanies his experimental film opera, *The River of Fundament* (completed in 2014 in collaboration with composer Jonathan Bepler), by giant sculptoid forms in crushed automotive metal and other found material that restage images from its visionary master text, Norman Mailer's 1983 metahistorical epic on Egyptian archetypes called *Ancient Evenings*. This is a lengthy mass-market novel identified within the exhibit, in several vitrine sculptures, as the "libretto" for the opera. In one such piece, called *Ancient Evenings: Ka libretto* (originally constructed at the inception of the project in 2007), the composite *objet* is rendered extra legible one might say, in my present "reading," within a networked analytic circuit along with the Middle Eastern ironies of Forte's *Desert Parcel.* For here the Mailer novel is palpably "abridged" (as it would have to be, if otherwise, for filming) by the sawing out of a middle fourth of its vertical bulk. The remaining top and bottom segments—opening a stained, striated crevice between them, with their upper page surface sketched upon as if in storyboard images from the pending film—are further aligned with, and cradled by, two other canonical American bestsellers flanged out (upside down to each other) as a kind of lectern.

But that is not the whole of it. All three books are in turn supported in an angled carved pedestal of salt, no doubt alluding to the chemical basis for Eyptian rites of mummification. Here, then, and as if with a further cyclic hint of the elegiac redeemed by resuscitation across the metaphorics of the two Hemingway titles, *For Whom the Bell Tolls* and *The Sun Also Rises*, a "Mailer" is thus preserved for the ages on the shoulders of his macho predecessor as well as in the embalming chemistry of the cult practice his own novel celebrates and reworks. As with so many *bibliobjets*, there is little of medium-specificity in these codex works by Forte and Barney, *Ancient Evenings* and *The Book of Maladies*, yet nonetheless a riot of medium-specification. The matter of one book is sliced, scarified, sketched upon, sacrificed to interpretation in the name of its own mythic theme, the other book's material surface stained and corroded on behalf of its diagnostic topic. To generalize: imprint is replaced by impress, whether overlaying the denatured page or further distressing it. To analyze: content has again succumbed to the newly figurative version of its own tampered form.

Then, too, in contrast to this book you cannot read, sawed up, drawn upon, and sealed under glass for museum display, there are the canvases you can do nothing with but read. In this way, the *bibliobjet*, which figures the book by disfeaturing and so defeating all legibility, is the conceptualist opposite number to what I have earlier investigated as the "lexigraph" on wall as well as canvas (Stewart 2007: 329–74). One particularly witty and involuted version of this appears in a 1974 work by Los Angeles artist Allen Ruppersberg subsequently acquired and recently displayed, four decades later, by the Museum of Modern Art in its 2014 New Acquistions show under the gallery rubric "Sites of Reason"—rather than sights of images: a curatorial foregrounding of the relevance to reception of precisely the works' localized analytic network. Certainly no art could be less sight-heavy or pictorial than Ruppersberg's *The Picture of Dorian Gray*, a work of "marker on canvas" that writes out in bold and indelible cursive hand every word of Oscar Wilde's ekphrasitc novel depicting the soul-searing effects of a magic portrait. Seldom has sustained museum contemplation been more strenuously invited, since to make one's way along and (with more difficulty) back across the thousands of lines of cramped text, the reader, as Ruppersberg explains, would, in the inordinate time commitment required by this "work," have been made to suffer at least some of the aging process that terrifies the novel's vain hero, though registered in his case only on the canvas of his degenerating portrait. The story's black magic is thus re-displaced, as it were, from these fixed and unchanging "text blocks" leaned against the gallery wall (of the canvas size required for a full-length portrait) even as it is normalized in the time-based medium of narration per se, which may enchant but also ages the reader in the very grips of its times-stands-still fantasy.

It is no accident that this is a work from the same Allen Ruppersberg who, also in the 1970s, did line drawings of famous books captioned with their "reading time" etched across the bottom, as if they were prerecorded audiobooks instead—or as if the medium of phonetic language could in this madcap way be routinized, quantified, and assigned. The mere picture of a book you cannot in fact spend time reading versus the canvas sequence about the eerie picture you cannot see but about which you can only read over time: two sides of the same demediating coin. Each is concerned with aesthetic substrates that both bracket a conceptual gambit and bridge an aesthetic dualism older even than the transmedial labors of ekphrasis, with its impulse to paint in words. For, beyond this, they carry us back to the very difference between temporal and plastic art. That return can take even more curious forms when one shifts, now, from the handmade to the electronic spectrum of new conceptual work: the latter concerned with the word/image divide in certain overriding phono/graphic transmediations.

## 3D printing and the e-lided text

Digitization has itself upped the stakes of ironic distancing in this line, and the previous book anatomist, Fiona Banner, has recently spoofed the faux 3D mock-up of remediation in formats like that of Google Books by showing us, not the "Look Inside" function implicit by graphic dog-eared page (or explicit on Amazon's website), but a real book one still cannot ogle for its cover image even when holding it in one's hand. This is her lampoon of the pageless scroll of backlit images in the e-text sampler: accomplished by photographing a 3D yellow book whose otherwise blank cover, in actual red lettering, reads "No Image Available"—as if in some default of the original designer rather than the computer engineers. But beyond travestying in this way the computerized "reissue" of the absented codex, there are other more comprehensive rewritings, it might be said, of print culture by electronics that bear consideration in closing, given their radical breach of the eponymous media|matter divide.

   I am referring here to something more technologically advanced than mere technical irony. This is because the relation of the codex form to the digital ascendancy has yet more abstract and oblique ramifications than just in such comedies of "transposition" (as Henri Bergson would call them) where a work in one medium—as I would rephrase his logic—can only be understood in its transit to (and transaction with) another (Stewart 2011: 85–7): for instance, text into sculpture and back again to metaphor. Instances proliferate of electronic text being brought into the conceptual orbit of codex bookwork: the burned I-pad rather than verboten book, the print volume flattened and carved into an open laptop, complete with raised keypad and CD/DVD drawer awaiting its own new-media inserts (including several variants of such sculptural ironies). Or note, in the work of Evan Roth, an entirely new valence for the "portrait of a reader" (long-standing topos and commercial mainstay in Western painting), which takes the form in Roth's work of the portrait-scaled acrylic rendering of a thumbstreak (the size of a portrayed cranium) as a swipe-to-open biometric ID on the otherwise blank surface of an e-tool. But what if a supposed medium, or medial process, let us say print(ing) itself—that second and definitive phase of codex production for modern culture after the manuscript era—no longer quite says what it means as a term? If print has lost recognition as such, as the laying down of lettered text, what about reading? It is again my premise, to be tested on this latest extreme evidence, that reading (in the form of figural decoding) outstrips the mediamatter that would ordinarily be thought necessary to sustain it.

After more than a half-millennium reign for the printing press, the two words that denominate its prolonged and signal contribution, even though ambiguous—designating either the ink and paper of impress or the meaning thus sheeted—have become, as italicized in a 2014 museum show, not a phrase any longer but a tacit clause: an infinitive or an imperative, with no end in sight. As impossible as it would have sounded until very recently, that is, the new engineering goal, beyond all imageering, is to *print matter* rather than just text. Few technological functions in our current computerized climate are more disorientingly transmedial—its nomenclature for starters—than such digital facilitations in the mode of "3D printing." Nor more clearly marked as such than when several strictly aesthetic (even conceptual, rather than practical) variants are placed on display together in an extensive 2014 exhibition at the Museum of Art and Design in New York City—with its punning idiomatic evocation of the imaginatively unmanageable (as well as the postmanual) in the rubric "Out of Hand: Materializing the Postdigital." Far afield from the raised lettering of braille, this dimensional printing is the newly computerized generation not of machinic signifiers but of incarnate signifieds, the triumph not so much of mind over matter, or even of sign over substance, as of signal over substrate: call it the coded manifestation of spatial form. Printing? Digital spray painting is more like it, in a polymer impasto thick enough to count as digital sculpture—or in one remarkable transmedial (audiovisual) leap, digital pottery.

Far short of such a cutting-edge exhibition, print matter is now closer to print "smatter," no longer the result of cold type but of laserjet spray. Yet digital printing is one thing, digital engineering another. In widening (indeed exponentially expanding) the technical parameters for the logic of demediation in regard to the codex form (and its resultant transmediation), one can now entirely bypass the lineage of Gutenberg—or say stretch it beyond recognition altogether—in the conceptual rather than merely instrumental ventures of this so-called 3D printing. There at that MAD show (pun intended, as many of its own objects would tend to suggest), one exemplary work by François Brument stands out for the way it defeats categories by intimately transfusing them. For beyond anything before it in either form, this synthesis of speech pattern and sculpture—well beyond lexigraph or *bibliobjet*—intervenes at a wry and retooled angle in a technology that ordinarily operates by eliminating any vestige of the linguistic, let alone the vocal, in the strict mechanics of the no-longer-verbal printing process. But in Brument's adaptation of the 3D apparatus, it is still words, or at least the phonic rudiments of speech sounds, that all but literally see their way to print, yet this time without a typographic surface of their own. They are manifested only as what they physically (rather than semantically) conjure.

This takes some further explanation, of course, as well as some standing back for context in that whole high-tech (that hyper-tech) exhibition. For as if emblemizing the residual or anachronistic term "printing" most directly of all the objects and installations on view there, Brument's invited sounds (from gallery visitors) are not transferred to ordinary printed matter but bypass that kind of imprint altogether in their transit, via wavelength graphics, to sculptural materialization. His apparatus—and its resultant polymer constructs—thus implicitly loop back through the very birth of humanism with Gutenberg to route us round instead to the limit of discursivity itself in the ultimate medial passage (wholly computational now) from idea through plan to manifestation, unimpeded by the language of human directives. Put otherwise, on exhibit there at MAD, in this single strange project, is the general tenor of the entire exhibit: the wholesale implosion of both design and art alike into the totalized computer program and its quick-dry extrusions, rescued for reflection only by way, in this case, of a transmedial conceptualism underlying the sheer display, one capable of a tacit punning matrix in the new ironies of so-called plastic art.

## InterVase: The vanished material interface

As Brument's work passes through certain resonances with aesthetic history, one can think of it under the banner of ex-phrasis. All actual phrasing in the classic mode of ekphrasis—verbal art representing visual form (think Keats's *Ode on a Grecian Urn*)—has disappeared here into a voice without referential input, a sound without speech or sense, that can nonetheless, as it were, declaim shape (full explanation still pending, I realize.). Whereas the typical 3D materialization on offer at MAD is generated from a computerized optic scan of object or body in the first place, then spewed forth in duplication, Brument's rarefied instance of such printing in *Vase #44* (2009) is a displaced form of audial-recognition that brings a potentially eclipsed materiality back to the prosthetic powers of an enunciating (rather than seen and laser-traced) body. It does so, one might say, in the bizarre electronic update of a dictograph from the preceding century. Or think of it this way: voice-recognition software taken to a new digital extreme in voice-imaging, so that what is made visible from human sound is no longer an array of signifying letters but an arbitrary palpable shape. Where sound waves once merely displaced the atmosphere in spatial traverse, now they can make things *take place* out of thin air. What thus results from the audiovisual participant's recorded voice, wordless or otherwise—once engaging the mic and pressing the transmit button— is not a glass-blown speech bubble, for instance, but rather a digitally fabricated and freestanding piece of hollowed-out form in polymer build-up: a vase whose angles, shape, volume, and conical breadth constitute the 3D translation of pitch, volume, duration, and the rest, in the enunciating utterance. Fiat form; let there be matter: the Logos as Wired Word.

With Keats's Urn, there in the British Museum rather than at MAD, the persona's troping of the ancient vase's figuration in various ways—vivifying its sculpted legends along a projected emotional and phonological trajectory matched to a 360-degree inspection of the material object itself—amounts almost to turning the artifact in the poet's own refashioning hand, all that euphony and metaphor at its fingertips. With Brument's *Vase*, in contrast, enough guttural *urs* may turn into an urn through merely the least grunt of sub-verse intonation. The rhetorically shaped artifact and its rounded internal double in the *Ode* has become, two centuries later, simply (if astonishingly) the signal-generated receptor and its microfashioned receptacle. So that "please do not touch" has been outplayed by the automatically push- button manifestation of a once hand-tooled craft. For whereas Keats could only call out to the permanence and visual authority of an antique art in the supposed half measure of "word pictures," here, under the new digitized mediation of voice transcription, any museum goer, however talentless with words or hands, can, by throwing his voice—it would seem almost the punning matrix of the entire work—thus *throw* a pot, that is, render a vase in polymer replica. Interface/vase: a programmed transit whereby, as in all these works at MAD—even when far less obliquely purposed and conceptual than this of Brument's—we find software operating to transform image into a new hardware, a new utensil or *objet* all its own. And another transmedial pun may seem actualized, literalized, in Brument's *vase*: sonic waves manifested, as if via another idiom made material, through the rippled and folded surfaces of a digital urn whose own spatial *volume* is only the materialization of an almost palpable sonic one. In this transmedial throwing of the voice, ventriloquism has become *ventriloptic* in a new modality of plastic art. Nothing could be more, as it were, purely conceptual in its materialization except perhaps, by a kind of verbal irony of concept itself, the piece of furniture—the furniture piece—located at the other side of this gallery from Brument's voice sculpture: a wildly corrugated brain wave sofa contoured to the EKG patterns accompanying, as graphic template, the very thought of rest and relaxation. Mind over matter indeed: interior décor gone freakishly far inward.

But why, finally, all those visual puns—or optical allusions—lurking in the premises of such display at its most conceptual? For the same reason, I find, that the mute and sometimes mutilated bookwork—its bankrupt and reinvested medial status coming through to us only by what Dworkin calls "analysis"—seems to spell forth its foreclosure or abasement to the interpreting ear of the viewer. For such is a spectator undaunted by illegibility in a desire to read the book form. Or in the case of Brument's work, to read the worldless work of human voice. The lectoral, in Roland Barthes sense, is the vector of attention, we might insist, even in works he once called *illisible* (Stewart 2007: 68–70), by which he was referring to the teasing phantom alphabets mastered by Michaux or Saul Steinberg (or even, one finds, Picasso) that Barthes himself also experimented with. But by the notion of lectorship I would mean, more broadly, those various instances, sculptural as well as lexigraphic, in which the shape of textual mediation, depurposed and turned (inward) upon itself, transmediates its own former (or still latent) verbal purport into a different tropological system. This is where meaning still borrows from the literary (and its wordplay) in the absence of script or grammar, making even spewed polymer art silently articulate by way of retinal metaphor. To reiterate my opening sense of the place of all such verbally demediated conceptual works in the considerations of this volume, any such disclosure of *mediamatter* specifies itself in the singular (*medium matter*), or sometimes in its discernibly hybrid forms, only so as to perform its deeper work of intersensory transactions and cross-cognitive ironies. It is at this point that what you see is not what you get, but only its optic instigations in the grip of response, involving as it does reception's not just discerning but frequently discursive eye.

Now, imagine ancient Greek lettering carved into the semicircular bulk of a crumbling funereal urn in cut-paper simulation hewn from fanned-out multiple paperback volumes of English romantic verse, the truncated sculptural form resting in place, and in peace, on a marble lectern over the title *Ode on a Grecian Urn (after Keats)*. Bookwork par excellence in the transfer of unreadable content into legible form. But now this: the vase not even laserjetted into bas relief on a page but sprayed into autonomous form by sheer sublingual voicing. Instead of the Keatsian "O Attic shape!"—bolstered by the poem's lexical alchemy of summoned image—just sheer noise. Again, then, I stress the span of consideration entitled by this essay in its negotiations "between print matter and page matter." For in its extreme elision of this difference in Brument's electronic installation, as if in a parody of the logocentric *Voice* as the issuing source of all script, the *Vase* resulting from the immaterial interface of sound waves and quasi-holographic sine curves is only the most technologically enhanced form of the conceptual sculptures—or say sculptural concept—under broader revisitation here (since *Bookwork*). It is a case of legibility disappearing in deference (not reference) to an alternate objectification of its shaping force in the mind's eye.

That everything the presswork of "printing" has for centuries seemed prompting readers to hallucinate into cognitive presence from mere letters should now, in the new millennium, be a name for an actual 3D version of digital impress and build-up—this is a phenomenon,

I have wanted to show, that has opened not just a new commercial avenue for industrial fabrication but yet another new corner gallery in postmodernity's conceptual museum. The museum in this regard, however, though long harboring its own protective space for the "mediatic system" (Jameson again), no longer offers so securely a place apart, "utopian" or otherwise (in his skeptical terms). For the aesthetic "network" through which the media object "analyzes" itself into definition (Dworkin once more), in the less fringe than futurist case of Brument's *Vase* work, approaches the point where it stands revealed as nothing more, nor less, than another subordinate digitized function of the blanket webworked Internet through which the weightless bulk of our signage, and now even the substantialized signals of our sculpting, can increasingly perform their remote-controlled tasks.

# References

Butor, M. (1996), "The Book as Object," in *Inventory: Essays by Michel Butor*, ed. and trans. Richard Howard, New York: George Braziller.

Dworkin, C. (2013), *No Medium*, Cambridge, MA: The MIT Press.

Foster, H. (2013), *The Art-Architecture Complex*, New York: Verso.

Jameson, F. (1991), "Space" ("Utopianism after the End of Utopia"), in *Postmodernism, or the Cultural Logic of Late Capitalism*, 154–80, Durham, NC: Duke University Press.

Lurz, J. (2010), "Mediation and the Object of the Book," *Critical Inquiry* 36 (Spring): 348–54.

Mitchell, W. J. T. (2005), "There are No Visual Media," *Journal of Visual Culture*, 4 (2 August): 257–66.

Stewart, G. (2007), *The Look of Reading: Book, Painting, Text*, Chicago: University of Chicago Press.

Stewart, G. (2010), "Bookwork as Demediation," *Critical Inquiry* 36 (Spring): 410–57.

Stewart, G. (2011), *Bookwork: Medium to Object to Concept to Art*, Chicago: University of Chicago Press.

# 4

# "Local Color": Light in Faulkner

### Hanjo Berressem

**Unidentified participant**: *Sir, I've been told that the title of* Light in August *came from a colloquialism for the completion of a pregnancy. Is that true?*

**William Faulkner**: *No, I used it because in my country in August there's a—a—a peculiar quality to light, and that's what that title means. It has in a—a sense nothing to with the book at all, the story at all.* (Faulkner, "Interview")

## Introduction

In literary studies handbooks, "local color writing" is usually described as a variation of nineteenth-century realist fiction that portrays specific regions and landscapes, with a particular focus on a verisimilitude of dialect, custom, and geography. Most of it is set in rural surroundings; small villages or towns that are inhabited by the proverbial common people or, to capture its prevalent mood more concisely, by "plain folk." In fact, a large amount of local color writing is not only provincial and pedestrian in terms of geography but also—and for good reason!—in terms of the poetics, politics, and values it advocates.

It would be wrong, however, to reduce local color writing to this "light-weight" version. In fact, the writers who have influenced and shaped our image of local color writing are among the most cherished of American authors, such as Bret Harte, Mark Twain, Sherwood Anderson, John Steinbeck, or Kate Chopin. Between these, the spectrum runs from Harte's cold gold-rush realism, Twain's darkly humorous satires about sweaty Hannibal, Anderson's Winesburg main-street emptiness, Steinbeck's dry and dust-bowled fields, to Chopin's adulterous, sun-drenched islands and beaches.

In the twentieth century, "local color writing" refracts into primitivism, agrarianism, provincialism, and, perhaps most generally, regionalism. It also enters a media ecology that includes as one of its important elements documentary writing, photography, and film—as Lothar Hönninghausen notes, one can easily draw a line from the "slow overalls" in William Faulkner's *Sanctuary* to the "over-hauls" in Walker Evans and James Agee's *Let Us Now Praise Famous Men*—as well as painting, theater, and music.

One author who is missing from the above list of "high-end" local color writers is William Faulkner, who produced some of the most complicated, sophisticated, and fascinating regionalist literature in twentieth-century America. At the same time, his work does not easily fit many of the local color stereotypes. Not only does his writing evade and at the same time transcend what are generally considered prevalent local color sentiments, but it is also not carried by the common, realist narrative voice of local color writing; a voice that oscillates, by default, between caricature, tall-tale humor, melodrama, and social criticism. Part of this is owed to the fact that Faulkner's prose is refracted through modernism—most intensely, perhaps, in the experimental narrative architecture of *The Sound and the Fury*—the other to his obsession with the "unmediated" portrayal of character and psychology through the faithful reproduction of thought by way of stream of consciousness and interior monologue.

In both its complexity and singularity, Faulkner's work allows expanding the term local color toward a more contemporary notion of what it means to develop a story from within a specific, concrete environment. Maybe Faulkner's writing is less "local color" or "regionalist," in fact, than a form of "literary environmentalism": an ecological writing without the ethical overtones nowadays attributed to "the ecological" as a political program.

Like Anderson's Winesburg, Ohio, Faulkner's Yoknapatawph County is a fictional landscape that shares all of the givens and characteristics of a specific geographical "postage-stamp"—which is the term Faulkner used to show how microscopic he considered his "environment" to be—of a real landscape, except for the fact that it is not real. Its status, as "quasi-real"—Faulkner describes it "as if" it were real—calls up a main characteristic of local color writing, which is that the writer should have a firsthand experience of living in the geographical region the story is set in. A local color poetics calls not only for an intimate knowledge about a specific region; this intimacy presupposes a life, or at least a good amount of time, spent in the milieu that is being described. A local color poetics, then, demands both an intimate and an immediate experience of its sujet. It would be very unusual to write a local color story about a region one has not really lived in, although it could of course be done, for instance, by way of what Thomas Pynchon has called relentless "Baedeckerizing."

The story, or perhaps "urban myth," of Faulkner's reaction to writing screenplays in Hollywood on the premises of the 20th Century Fox studio, can be read on this background: not having written anything for a while, he told a producer that, although he had an idea for a story, he would be better able to write it at home rather than in the Writers Building. The producer assumed that Faulkner meant his home in Hollywood. A few days later, however, he got a call from Faulkner, who had returned to Oxford, Mississippi.

In biological terms, writing a successful local color story calls for what Humberto Maturana and Francisco Varela call a long-term "structural coupling" of the writer to a specific milieu and the interaction of its elements. As Maturana and Varela note, "[e]nvironment and unity act as mutual sources of perturbation, triggering changes of state. We have called this ongoing process 'structural coupling'" (Maturana and Varela 1998: 99). The infinite number of these interactions form "a network of multidimensional structural coupling. Indeed, living systems and their conditions of living, whichever these may be, exist in a network of continuous structural coupling, and change together congruently in a process that spontaneously lasts as long as the autopoietic organization of the living systems is conserved" (Maturana, "Autopoiesis").

The reason I differentiate between "interactions" and "relations" is that I take the former to pertain to life as enacted, while the latter pertains to life as observed. While what Maturana and Varela call "the observer" might well write very insightful literature, he cannot write successful local color stories: The local color writer is immanent to the milieu, while the observer looks at the milieu from the outside—which is not quite true, of course, because in the best of possible worlds the local color writer is both immanent to a specific milieu and at the same time a good observer of that milieu. Every human being, in fact, shares this double determination of being both a living entity, which means a being that is defined in parameters that are "purely zoological" (Maturana 1994: 80) and which exists in a space of unrelational irritations, and of being a *Homo sapiens sapiens*" (80) who exists in a "relational space" (80). A third element of successful "local color writing" is, of course, that one needs to be a good writer. While "true" local color writing never lacks in the first demand, it often does in the other two. Or, to be less judgmental, in Faulkner's case, these three demands are all met. The unnamed narrator in "A Rose for Emily," whose voice remains hidden behind the collective "we" of the villagers, is a good example of Faulkner's complex "imagining" of an anonymous observer.

In this context, Yoknapatawph County's status is both utterly real as "lived in" *and* utterly fictional as "observed." While Faulkner as a living being is immanent to the factual site, in terms of the fictional site, he is—like everybody else who lives in it—its observer and as such its creator in the sense that every observation is inherently fictional. Additionally, however, he also creates it, as a writer, within the milieu of "the literary." Given these complex layerings, local color writing might be defined as fiction that integrates structural couplings into the act of writing; as fiction that is written from within the field of structural couplings; translated into aesthetic registers: it is a writing that not only represents a specific site but also expresses it.

Except perhaps in the genres of "life writing" and "local color writing," and despite the rise of material culture studies and the various forms of new materialism, literary studies still tends to look at literature as uncoupled from the life of the author, from natural processes and from the notion of "life" in general. In fact, when it is not derided as downright naïve, the level of the immediate immersion *in*, and experience *of*, the milieu still tends to be neglected in literary studies in terms of both theory and practice.

To a time that has witnessed a long history of the fear of essentialisms—which has been the main reason for the "repression" of the fields of biography, nature, and life—Faulkner's work, in which the notion of a deep immersion in a milieu and an acknowledgment of the agency of that milieu is so "naturally" given, must come across almost as a provocation. In all of Faulkner's work, in fact, the characters'—as well as the writer's—actions and thoughts emerge, invariably and inescapably, from within a specific milieu. "A man would have to act as the land where he was born had trained him to act" (Faulkner 1981: 192), Joanna Burden tells Joe Christmas in *Light in August*. And when the narrative focalizes Christmas a short while later, he notes that "with the corruption which she [Joanna] seemed *to gather from the air itself*, she began to corrupt him" (195, my emphasis). Similarly, in the voices that surround him, Hightower hears "his own history, his own land, his own *environed blood*" (276, emphasis added).

While one might argue that these are merely voices within the narration that call for detailed deconstructions, the narrator does not restrict this logic of immanence to human beings. In fact, he quite programmatically includes resonances of nonhuman agents and nonhuman thoughts with the milieu. The wagon in which Mr. Armstid takes Lena to Jefferson, for instance, "has a kind of rhythm, its ungreased and outraged wood one with the slow afternoon, the road, the heat" (12). Not only the humans, but also the nonhumans in the scene, such as the "outraged wood," are pervaded with both physical and psychic affects that are caused by and enacted within the milieu.

What saves such descriptions of immanence from advocating an essentialism is that they are not so much about fixed substances as they are about habits and the processes of the contraction and the dissolution of these habits; of doing habitual things as well as things that are "out of habit." This getting into and getting out of habits pertains once more to both the land and to its inhabitants, which are indeed "structurally coupled," forming a milieu in which each element forms and is simultaneously formed by the respective environment. When the county changes, the people change. When the people change, that changes the county.

From such a habitual perspective, nothing is ever at rest in an environment; something to which the many journeys and movements in Faulkner's plots attest, all of which trace specific people moving through a milieu and the changes they, both consciously and unconsciously, cause in it. You can take the people out of Yoknapatawph County, one might say, but you cannot take Yoknapatawph County out of the people. At least, any such extraction will take a very long time and much effort. Not only does the county cling to the protagonists like the proverbial mud and *vice versa*; there is also, for some characters, a strong desire to become and to remain rooted, such as Thomas Sutpen in *Absalom, Absalom!* Whether the characters are settled or not, however, the county has molded them into what they are by way of long routines of living, feeling, and thinking. Both their bodies and their thoughts express the county in the same way that they are expressions of the county.

Of course one might argue that, on this background, the most radical way to expand the field of local color writing would be to consider all of literature as local color writing, in the sense that all of literature emerges from site- and time-specific situations and all writers are, whether they like it or not, living observers. According to this at first sight counterintuitive logic, even writing that seems to have nothing to do with local color writing would be considered as local color writing. While I would in some aspects consider this a valid approach, it is, at the same time, counterproductive to expand a critical term to such a degree that it comes to dilute all possibilities of differentiation. I am not promoting, therefore, a globalized notion that considers all of literature as "categorically" local color writing. In fact, given its stress on the local, such a globalization would be especially ironic in the case of local color writing: there are critical differences between, say, "experimental writing" and "local color writing" in the same way that there are different degrees of local color writing. Also, one needs to take into account the history of the term. Most of all, however, the challenge is to look at singular writers and singular works of literature; to see how they negotiate the field of local color, because everything in the local color universe is a singularity. In my case, this will be Faulkner's novel *Light in August*. Before I get there, however, some general remarks.

## The elements of Faulkner's South

One might say that a heading such as "The Elements of Faulkner's South" asks for at least a full monograph. All I want to point out in this essay, however, is Faulkner's singularity. The well-known, but nevertheless seminal fact that Faulkner's novels are set in both an extremely fictional and an extremely real landscape that is neither the cliché South of romance, with its technicolor plantations such as *Tara* or *Belle Reve*, nor the documentary South of social statistics, with its cultural and historical factualities. How does Faulkner succeed to create a South that is radically fictional and singular, while it is, at the same time, radically factual and "universal"?

The reason, I think, lies in the way Faulkner molds the stories from within the "elements" that make up the South. As I noted, these elements are not the default ingredients of a South constructed from a history of French aristocracy, an economy of slavery and cotton, and a sociology of chivalry and cruelty. While these are important cultural and historical elements, the more "elementary elements" from which Faulkner develops "his" South show that these cultural elements are in turn immanent *to* and emerge *from* a more *elemental*, natural landscape in the sense of the "natural elements" earth, water, air, and fire, translated into Faulknerian terms: farming, rain and floods, storms, heat and draughts. The constantly changing mixtures of these natural elements, which one might further correlate with the four states of matter—solid, liquid, gas, and plasma—make up the "originary milieu" and the climate to which Faulkner's stories are immanent. They are the material from which they are built rather than their background coloration.

This is why Faulkner is, perhaps, more of a naturalist than a realist. As Gilles Deleuze notes, while realism is concerned with "derived milieus" naturalism shows how derived milieus emerge from "originary milieus" or, more often, how they "sink back" into them. As Deleuze notes *à propos* Émile Zola: "He had the idea of making real milieux run in parallel with originary worlds. ... This is the essential point; the two will not let themselves be separated and do not take on distinct form" (Deleuze 1986: 124). The originary world "only exists and operates in the depths of a real milieu, whose violence and cruelty it reveals. But at the same time the milieu only presents itself as real in its immanence in the originary world, it has the status of a 'derived milieu', which receives a temporality as destiny from the originary world" (125). The originary world "does not exist independently of the determinate milieux, but conversely makes them exist with characteristics and features which come from above, or rather, from a still more terrible depth. ... Milieux constantly emerge from the originary world and retreat into it" (125–6).

If I talk about the elements that make up an originary milieu as "media," this means that my use of the term "media" does not refer to the default notion of media as man-made, technical media of information, such as the telephone, the radio, film, or the computer. Rather, it refers to the notion of media prevalent in the natural sciences, where they designate habitats or environments, such as water for fish, air for humans, or a nutrient solution in a petri dish for bacteria. Structurally, these media consist of a multitude of small, loosely coupled elements from which larger, more complex and more strictly coupled ensembles are created. These loosely coupled elements form the building blocks of both the material as well as the immaterial world (and as such they are also the media from which the technical media are assembled). Drops of water are the media of a river in the same way that molecules are the media of living bodies.

A passage from *Absalom, Absalom!* illustrates how fundamentally important these media and their molecular structure are for Faulkner, who expresses most of what I have said until now in a much more poetic and concise manner. The direct context is Quentin contemplating the relation between himself and his father:

> *Maybe nothing ever happens once and is finished. Maybe happen is never once but like ripples maybe on water after the pebble sinks, the ripples moving on, spreading, the pool attached by a narrow umbilical water-cord to the next pool which the first pool feeds, has fed, did feed, let this second pool contain a different temperature of water, a different molecularity of having seen, felt, remembered, reflect in a different tone the infinite unchanging sky, it doesn't matter: that pebble's watery echo whose fall it did not even see moves across its surface too at the original ripple-space, to the old ineradicable rhythm. ...* (Faulkner 1982: 210)

The history of the notion of the four elements as media goes back to the Pre-Socratic Philosophy of nature developed by philosophers such as Heraclitus and Empedocles, and is taken up by Plato and Aristotle. It might be worthwhile, in fact, to analyze Faulkner's work according to the specific element that is foregrounded in each one of them, such as water in *As I Lay Dying* and *The Old Man* (the passage through the river, the flood), the earth in *Absalom, Absalom!* (Sutpen buying land from the Indians and building "Sutpen's Hundred"), fire in *Light in August* (Joanna Burden's house burning down), or wind and air in *Wild Palms* (the drafty cold of the Utah mining camp). Although it is intriguing, such a typology would be too restrictive, however, if it were used as more than providing the "tenor" of the specific text. As they form multiplicities, all of the elements—as well as the specific channels of perception that go with it, such as hearing in *The Sound and the Fury*, smell and taste in *Absalom, Absalom!*, vision in *Light in August*, touch in *Sanctuary*—are present in all of the novels, such as "fire" and "water" in *The Sound and the Fury*. Rather than aim for a tendential typology, therefore, let me follow in more detail "light," the "fifth" element that suffuses all novels to a similar degree.

# Light

In his theory of the elements, Aristotle added the "lumiferous aether" to the four elements as the "fifth element" that brings the other elements to life and that is, at the same time, the optical medium in which all of the other elements are immersed. Light, as the visible aspect of the electromagnetic field that "is" the world, is made up of small, loosely assembled elements—photons that show the characteristics of both particle and wave and that are, in that complementarity, both singular and communal—and it is a natural habitat in the sense that every landscape is suffused with light.

In terms of media studies, one of the earliest descriptions of light as a medium comes from the essay "Object and Medium" (*Ding und Medium*) by Fritz Heider. Heider treats light and lighted space as media of perception (*Wahrnehmungsmedium*). "Mostly, the objects of perception are the solid and semisolid objects in our surroundings, and the mediator is the air-filled space, the medium that surrounds the objects" (Heider 2004: 322), he notes. The optical medium consists of the "multiplicity of light-waves" (323)—the "photonic multiplicity"— within which images of objects move.

Heider's basic idea is that "the specific characteristic of the medium" consists precisely in being "to a large degree unimportant for the form of what happens" (324). As a loosely coupled set of elements, it is a medium that invariably takes on the specific "eigenform of the substrate" (323) in which it actualizes itself. In fact, unlike objects within it, the medium of light does not have an "eigenoscillation" (324) and is, therefore, not a coherent "unity" (325). Its only characteristic, in fact, is to be a "substrate without characteristics." It is a pure multiplicity made up of *"many independent parts"* (326). Light rays, for instance, "are arranged in an atomistic juxtaposition" (332–3). The difference between "eigenhood" and "uneigenhood" is the categorical difference between forms and media. "The objects contain the real eigenhappening, the eigenoscillation" (329), Heider notes, while the medium can only be formed "into false unities, the oscillations forces upon them" (329). Quite paradoxically, then, according to Heider, the only characteristic of a formational medium is to be uncharacteristic. It functions merely as a substrate for forms or bodies. The form incarnates itself in a medium for a certain duration—a material form in the case of formational media, an immaterial form in the case of informational media—after which it loses this form again, leaving the medium free to be assembled into other forms.

Although Heider is seminal in defining light as a medium, his demand for its imperceptibility disregards that "the art of art" as well as "the art of life," lies precisely in finding modes of expression in which the characteristics of the medium are used *for* and integral *to* specific forms. The artistic challenge is to find modes to attribute media and forms, whether it is the hardness of the block of marble for the sculptor, the grain of charcoal or pigment for the painter, the body for the actor, and, as Roland Barthes has noted, the grain of the voice for the singer: If the surface of the vocal chords is one part of the medium of sound (the other being the air), the grain lies precisely in the fact that this surface is rough and that this roughness interferes directly in what would be called, from a position of information optimation, the ideal, perfectly clear, and transparent sound. It is the noise in the sound, however, that defines a singular voice. Every voice has a "local color." A voice without local color—timbre—is boring.

If one were to ask a sculptor about the essence of one of his or her figures, it would only be in the second instance a concept such as force, beauty, or pain. In the first instance, it would be marble, clay, or wood. Because the force would be a marbled force, the beauty a clayed beauty, and the pain a wooden pain. Similarly, in the cinema, the essence of a movie is not the more or less melodramatic story, but the light in which that melodrama is given. This light, in fact, does not only show the melodramas; cinema, as filmmakers as diverse as Douglas Sirk, Rainer Werner Fassbinder, and David Lynch have noted, is itself the melodrama of light. Every story emerges from, is told, and falls back into shades of darkness and light. Light does much more than just make a story visible. Every story, in fact, is quite literally immanent to a specific light, such as the "magic hour" of Los Angeles or the cold, harsh light of New York. Light is to cinema what marble, clay, or wood are to the sculptor. The cinema creates moving sculptures made purely from light.

If one takes this approach seriously, the medium is important down to its smallest, most unconscious levels. The way a subtle shift in the light can change a mood or a sound can bring a memory into consciousness. The way a shift in the electricity of the atmosphere can decide a long-standing historical conflict. The way the hardness in the wood can change the expression in a face. Heider does not consider this recursion into the imperceptibly small. Although he mentions the "micro-happenings of the molecules" (329), he notes that these micro-happenings are "not important for what happens on our scale" (329). In Faulkner's luminist, "local light" poetics, I will argue, they are of immense importance.

## Luminism

The term "luminism," which was coined by John I. H. Baur in the 1950s *á-propos* American landscape painting of the nineteenth century, describes a "genre" of painting—although the term might easily be extended to photography and film—that is, in terms of poetics and of *sujets*, in many aspects the analog of local color writing. Concerning its inherent poetics, luminism marks the moment when light is treated as the true medium of painting. The analogy also extends to its *sujets*—carefully observed and depicted scenes from everyday life—and its prevalent "colorful" mood. Luminism, which develops around paintings such as those of the Hudson River School, will shade into Impressionism, which is, although it is formally much more complex than "local color luminism," similarly concerned with objects given "immediately" *in* and *by* light. In terms of a painterly luminosity, in fact, one might construct a direct "line of light" from a baroque chiaroscuro to luminism—whether in America or in William Turner's light-drenched, post-Tambora England—and further to Impressionism.

The notion of an American light has always fascinated artists, especially if they had moved to America from Europe, where such a light was missing. The members of the Düsseldorf school of painting, for instance, considered the new light from the beginning as a local, specifically "American" light. Although luminist painting often depicts the wildness and the sublime grandeur of the American landscape in an equally grand light, the luminosity of America becomes truly "local" around the middle of the nineteenth century, when luminism takes on more minute, local color sujets, such as James Augustus Suydam's *Long Island* (1862, Private Collection) or George Caleb Bingham's *Fur Trappers Descending the Missouri*, (c. 1845, Metropolitan Museum of Art), which might be taken straight out of a Mark Twain novel.

In its depiction of tranquil landscapes, calm, reflective water and soft, hazy skies, much of luminist painting, like much of local color writing, highlights the natural elements. These elements, however, are all seen and depicted "under the given" of light, which provides the movement with its name and which functions as the medium that encompasses all of the others: Invariably, the four elements are bathed in the fifth element: light.

If luminism can be understood as "local color" painting, "local color writing" can be understood as inherently "luminist." If luminism is based on a complex play between color and light, one of its main "painterly" challenges is how to evoke the pure, immaterial luminosity of light by way of material, inherently colored and therefore "dark" pigment. In short, how to create (the "illusion" of) immaterial light by way of material pigment? This question is not only technical, however, but also cultural, because the notion of light changes in terms of the history of art, where light can be understood from within a metaphysics of light (such as the religious, literally "golden" background of Christian icons), a physics of light (such as the naturalist, "luminous realism" of a George de la Tour), or a metaphysics of color (such as the "abstract," pure white of the canvas in modern art). In *Über das Licht in der Malerei* Wolfgang Schöne traces these historical shifts by differentiating between "eigenlicht" (*Eigenlicht*), "lighting-light" (*Beleuchtungslicht*) and "eigencolor" (*Eigenfarbe*). Eigenlight, which defines the "light metaphysics" of medieval painting (Schöne 1954: 109), refers to the light that emanates from within a painting and whose luminosity radiates to the spectator, which is why Schöne also calls it "sending light." The eigenlight becomes most tangible in the "eigenshine" (*Eigenglanz*) (25) of the golden backgrounds that stand for the luminosity of a God who is "in a literal sense, light" (64).

Between the fifteenth and the eighteen centuries, this metaphysical logic of eigenlight is replaced by the physical logic of a lighting-light that can originate either the outside or the inside of a specific pictorial space and that lights that space, which is why Schöne also calls it "showing light" (119). This light can be differentiated once more into a natural, an artificial, or a sacral light, as well as into a light that is neither of these and which Schöne therefore calls "indifferent" (112) in relation to these three categories. Between the fifteenth and the eighteenth centuries, this indifferent light defines the passage from a "sacral" light to the natural and artificial light that, together, will define the nineteenth century.

If, from medieval times to the nineteenth century "color was understood as a function of light" (208), in modern art, "light becomes a function of color" (200). Although each specific "corporeal color" (*Körperfarbe*) (201) has its own value of lightness within a spectrum of darkness and brightness—an "eigenlight" (*Eigenhell*) and an "eigendark" (*Eigendunkel*) such as a bright yellow opposed to a dark brown—its material color (*"stoffliche" Buntheit*) is structurally set against an overall "immaterial luminosity" (*immaterielle Helle*) (203): a light blue chair might look darker than a dark blue chair if the immaterial luminosity around it is low.

In a twentieth-century painting, the importance of the "local color" of bodies increases above that of the surrounding "local light" because the framework is no longer a "given" naturalism that is based on a poetics of figuration, but rather an "abstract universe" that is defined by the eigenvalue of specific colors: in the passage from figuration to abstraction, eigencolor takes over from eigenlight. In general, however, all of the historical phases define interplays of color and light, the most general question being "in the history of Western painting, how do the eigenvalues and representational values of color relate to the artistic medium of light?" (205).

In terms of color-theory, the distinction between material color and immaterial luminosity—"local color" and "local light"—plays itself out between the two logics of "additive" and "subtractive" light. While additive light is defined by mixing light of two or more different colors, and is measured in terms of the addition of differently colored luminosities that together form white light, "subtractive light" is defined by subtracting parts of the spectrum of light that is "virtually present" in white light by means of colored pigments, and is measured in terms of rates of color and the absorption of light. All pigments or filters together lead to black.

The overall difference, then, is between virtual luminosity (light) and actual pigment (color); a difference that has been read in analogy to the fields of the body and the mind, respectively: Light is incarnated in a color "like" thought is incarnated in a body. In the more focused context of painting, the distinction has to do with the pigmented bodies of the canvas and the resulting colors on the one hand, and the virtual luminosity of the scene that is depicted on the other. On this background, luminism might be defined as the game of using the subtractive logic of colored pigment to simulate the additive logic of projected light.

Although the creation of the illusion of luminosity by way of colored pigment is an inherently painterly challenge, there are analogies to the "local color phenomenon" in literature. At the same time, however, there are critical differences, the most important of which is that the medium of language replaces pigment: the challenge of literature is to create the illusion of light through words. To understand that process, one needs to realize that the "local color phenomenon" I am trying to develop is precisely the opposite of what is called "local color" in the visual theory of color. In this theory, local color denotes specific colors of specific objects as *un*modified by ambient light and luminosity. It denotes the "pigmented color" of an object, without the interference of its position within an overall optical milieu. In the "local light phenomenon," in contrast, ambience and ambient light are the media that suffuse both the action and the objects that are being described: the ambient light suffuses the specific scene and, in extension, the whole world. Even more, it is the medium within which the action, the objects, the situation, and the world emerge. In luminist painting, every local color is suffused by a specific ambient light, as in the "proto-luminist" paintings by Nicolas Poussin or Claude Lorrain. Maybe one can sense the local color through the ambient light, but to posit a pure local color for the objects in these paintings is, ultimately, an idealization.

In fact, it is difficult to conceive, categorically, of the pure local color of an object, because that would imply the erasure of ambient light. Some modern museums attempt to approach such an ideal state by creating an ideally neutral architectural background and a similarly ideal neutral light in which to present the paintings; a light Schöne calls "site-light" (*Standortlicht*) (109). This is different in palaces such as the Palazzo Pitti or the Uffizi; pre-museal milieus in which the paintings shift with the day- and the night-light, without ever coming to rest in a milieu of an unchanging amount of a fully artificial light. Somewhat ironically, even luminist paintings are often suspended into such highly idealized visual milieus. Despite these attempts, however, the ambient light invariably interacts with the specific local color of the objects. As with the inhabitants of a specific milieu, local color and ambient light form complex optical mixtures.

## Deleuze: Light and the image

The awareness of these mixtures in his writing is what makes Gilles Deleuze talk of Faulkner as "literature's greatest 'luminist'" (Deleuze 2006: 68) in his book *Foucault*. The context of this remark is concerned precisely with the relation between the originary worlds of sound and light and the derived cultural milieus of meaning and vision; the way in which, in Foucault's work, individual knowledge and individual vision are implicated in the media of sound and light, respectively.

For Deleuze, the importance of Foucault lies in the way he problematizes the notions of inside and outside, but also in the way he thinks of sound and light as the two media to which the regimes of language and images are immanent. Foucault's work is literally pervaded with sound and light, Deleuze argues, and over long stretches, he traces in loving detail how in Foucault's thought, statements and images emerge from the media of sound and light, which in turn emerge from the plane of immanence considered as a "field of vectors" (7). "Just as statements are curves before they are phrases and propositions, so scenes are lines of light before they become contours and colors" (67), Deleuze notes. "The statement-curve integrates into language the intensity of the affects. ... But visibilities must then also integrate these in a completely different way, into light" (66).

Before they become meaningful, sound and light each form a "diagram of forces" (67) that is "realized" (67) or, in Deleuzian terminology "actualized", in "description-scenes and statement-curves" (67), respectively. Before individual knowledge and before an individual gaze "'there is' light, and 'there is' language"—"there is a 'there is' of light, a being of light or a light-being" (50). Benjy in *The Sound and the Fury* is perhaps the character nearest to this state of "there is language," but in fact all characters partake of this level, from the time-drenched language of Quentin to the dry, dusty, and violent language of Jason. All of these languages emerge from a sonorous ground; as do all visibilities, which are often related to the female characters, such as

Caddy Compson and "Miss Quentin" in *The Sound and the Fury*, or Temple Drake in *Sanctuary*.

As Deleuze states, "there is a 'there is' of light, a being of light or a light-being" (50). Everything that exists emerges from a luminous and sonorous multiplicity; a photonic plane of immanence:

> Visibilities are not forms of objects, nor even forms that would show up under light, but rather forms of luminosity which are created by the light itself and allow a thing or object to exist only as a flash, sparkle or shimmer. (45)

Already in "Michel Foucault's Main Concepts," which is in many ways a blueprint for *Foucault*, Deleuze claims that Foucault "paints the most beautiful paintings of light in philosophy and traces unprecedented curves of utterances" (*Regimes*: 260).

In *Foucault*, light is not only present in relation to philosophy, however, but also in relation to literature, as in the passage on Faulkner, and in relation to painting, where Deleuze finds similarly Foucauldian uses of light. Foucault's philosophical luminism has a correlative, for instance, in the luminism of Robert Delaunay about whose work Deleuze notes in *Abecedaire* under the letter "N for Neurology" that in it "light forms figures itself, figures formed by light." Delauney

> paints light figures, not aspects that light takes on when it meets an object. This is how Delaunay detaches himself from all objects, with the result of creating paintings without objects any longer. ... So in terms of the elimination of objects for rigid and geometric figures Delaunay substitutes figures of pure light.

All of these celebrations of light have ultimately to do with Deleuze's conceptualization of a luminous, photonic plane of immanence as the most comprehensive of all Deleuzian media. Photons are elementary particles that make up the sub-atomic building blocks of the world's electromagnetic field. They are the media of the world's "luminous reality;" a reality that is everywhere singular and different, because the light is everywhere and always different. In "The Brain is the Screen," Deleuze describes the various lights in terms of the cinema:

> You have one kind of light that presents a composite physical environment, and whose composition gives you white light, a Newtonian light that can be found in American cinema, and perhaps in Antonioni's films, though in a different way. Then you have a Goethe-light, an indestructible force that slams into shadows and picks things out. ... You have another kind of light defined by its contrast not with shadow, but with shades of white, opacity being a total white out. ... You have also a kind of light no longer defined either by composition or by contrast, but by alternation and the production of lunar figures. ... The list could go on forever, because new lighting events can always be created. ... (*Regimes*, 286)

Within this classification, every movie, as well as, in extension, every painting or novel is a singular luminous and sonorous event— image track and sound track—that has emerged from a photonic multiplicity.

## Local light

While it is quite natural that the visual arts should use light as its medium of choice, there is also a specific light and specific colors that define specific works of literature. In fact, it might make sense to talk of a "local light movement" instead of a "local color movement." After such a conceptual upgrade, the cliché of the "colorful characters" in local color stories is changed into something both more fundamental and local—into "luminous characters." Another characteristic of local color literature can also be "upgraded" in this way, although I will not deal with it in detail. If the treatment of language in local color is generally reduced to "the truthful rendering of dialect," a Deleuzian reading stresses how there is, to every environment, a specific inflection of "pure sound"; a "semiotics" that reaches from the low frequencies of a slow communal drawl to the high-pitched fever of a metropolitan conversation. It is on this level of pure sound that Hightower found his environed blood; a sound that is too faint to make out sentences and words—the anonymous sound of a specific community or of a singular human being such as Benjy.

It has taken me quite a while to get to Faulkner's *Light in August*. It is only after the various theoretizations of light and literature, however, that I feel I can adequately address Faulkner's use of light in that novel, whose "abstracted" original cover is a fitting illustration of its inherently luminous poetics.

In most of Faulkner's fiction, and certainly in *Light in August*, light is immediately related to a racial *chiarouscuro* that plays itself out not only in black and white, but, more importantly, in darkness and light as well. Given its topic, this is not surprising. In fact, it might at first sight seem to be something of a cliché. It is, however, a cliché only for certain figures within the narrative, who have come to feel and understand light and darkness as more or less fixed cultural markers by processes of slow habituation.

There are two remarkable passages in which darkness and light are the registers of the experience of raciality. In both of these, Faulkner treats the racially "derived milieu" of the South in terms of a more "originary milieu." The first passage is focalized through thirty-three-year-old Joe Christmas, who passes, culturally, as white, but who suspects that there is a certain amount of African American blood somewhere in his family tree. It never becomes clear whether that is true or not. He himself has only dark reminiscences of the children in the orphanage where he is raised calling him a "nigger." Most of his love affairs are attempts to find out if there is an elective affinity between him and African Americans; attempts that Faulkner describes as forms of a racial grisaille:

> He now lived as man and wife with a woman who resembled an ebony carving. At night he would lie in bed beside her sleepless, beginning to breathe deep and hard. He would do it deliberately, feeling, even watching, his white chest arch deeper and deeper within his ribcage,

trying to breathe into himself the dark odour, the dark and inscrutable thinking and being of negroes, with each suspiration trying to expel from himself the white blood and the white thinking and being. (Faulkner 1981: 170)

Throughout his life, Christmas tries to find out where in the spectrum between white and black he is situated. He does not do so, however, by reconstructing his past and his family tree, but by trying to see what happens when he *enters* or *moves* in a "dark milieu." In the intimate milieu of the bed, he attempts to literally breathe in a molecular blackness that he feels is exhumed by his black companion. The darkness at this point is not only a visual darkness, it refracts into three fields: a synesthetic, perceptual darkness ("dark odour"); an epistemological darkness ("dark thinking"); and an ontological darkness ("dark being"). The test is how and where he fits into this dark milieu, and what will happen to him when he literally breathes this milieu into himself.

A religious darkness is added to this racial grisaille(s) that is focalized through Joanna Burden:

A race doomed and cursed to be forever and ever a part of the white race's doom and curse for its sins. … The curse of every white child that ever was born and that ever will be born. … I had seen and known negroes since I could remember. I just looked at them as I did at rain, or furniture, or food or sleep. But after that I seemed to see them for the first time not as people, but as a thing, a shadow in which I lived, we lives, all white people, all other people. I thought of all the children coming forever and ever into the world, white, with the black shadow already falling upon them before they even drew breath. And I seemed to see the black shadow in the shape of a cross. (190)

The passage traces almost programmatically the shift from a perception of blackness as a natural element of the Southern milieu, to a cultural— in particular a religious—perception of blackness. In this series, African Americans are first experienced as "given," unquestioned elements in a milieu that reaches from the natural to the cultural in one chromatic

line: they are as common and as unremarkable as anything else, such
as "rain, furniture, food, and sleep." Although a first critical reflex
might be to read the comparison of African Americans to furniture as
an expression of racist dehumanization by Joanna Burden or even by
Faulkner, this would be too easy a reading. Rather, it is only after the
milieu of religion has entered their immediate and innocent immersion
into the general milieu that they become part of a religious allegory
that plays itself out no longer between white and black, but between
a sacral and allegorical darkness and a similarly sacral and allegorical
light. It is only when they are suffused by the light of religion that they
come to stand for one side of the polarity between light and dark; for
the shadow that falls on the doomed—the literally "burdened"—white
race. Similarly, the doctrine of Puritanism is distilled into Simon
McEachern's eyes, which "were lightcoloured, cold" (108). In the same
way that Nathaniel Hawthorne described the Puritan milieu as dark
and somber in the opening passage of *The Scarlet Letter*, opposing it to
the light, colorful milieu of the Native Americans, Faulkner stresses the
cold, inhuman light of the Puritan doctrine.

While such racial and religious grisaille(s) mark the naturalist,
violent luminism of the novel, other grisaille(s), which Faulkner creates
at a number of moments in the narrative, are not immediately religious
or racial. Some of them describe scenes that are just naturally luminous,
as when Christmas stops his horse at night, in a low, "shadowdappled"
(158) light, which makes the two look, before they start moving
again on the "moonblanched" (159) road, like "they might have been
an equestrian statue strayed from its pedestal and come to rest in an
attitude of ultimate exhaustion in a quiet and empty street and dappled
by moon shadows" (158).

In fact, there are many such moments of the meeting of darkness
and light not so much within the framework of a racial chiaroscuro, but
rather in terms of the meeting of a more elemental light and darkness
such as sunrises and sunsets; as when "after a time a light began to grow
beyond the hill, defining it" (82) or when "darkcaverneyed" (230) Gail
Hightower, who breathes "the rich maculate smell of the earth" (239),

sits at his study window waiting for nightfall; "that instant when all light has failed out of the sky and it would be night save for that faint light which day granaried leaf and grass blade reluctant suspire, making still a little light on earth though night itself has come" (46–7). It is not difficult to see in these gray chiaroscuros the dusky light of the romance and of dreams and visions of the past.

*Light in August* is not only about shades of gray, however. In fact, there is a very "colorful" quality to many of its descriptions of luminosity. Mrs. Armstid's face, for instance, when she watches Lena, has an "inwardlighted quality of tranquil and calm unreason" (16). While this is an internal, almost spiritual light, there are also many moments of a directly painterly, luminous suspension in a given milieu, as when Hightower remembers, truly *in* a moment when "the final copper light of afternoon fades" (350) and "the street beyond the low maples and the low signboard is prepared and empty, framed by the study window like a stage" (350), the time "when he was young, after he first came to Jefferson from the seminary *how that fading copper light would seem almost audible*, like a dying yellow fall of trumpets dying into an interval of silence and waiting" (350, emphasis added). At the end of this luminous reminiscence, "the copper light has completely gone now; the world hangs in a green suspension in colour and texture like light through coloured glass" (352).

The passage is filled with the potentialities of literature to address color; not only can it describe "interior lights" and call up specific connotations such as "copper," "like through coloured glass," or, in "the level, jonquilcoloured sun" (250), the color of the interior of the central cylindrical tubular projection of the jonquil flower. It can also do things with color that painting cannot, such as describing changes in light; its slow fading or its gradual change, but also a sudden change in light, as when the sky becomes, suddenly, a monstrous dark. It can provide the visual realm with a synesthetic quality, such as a sonorous quality when the "fading copper light" is described as "almost audible." It can put the quality of light at one or many remove and stress the materiality of color and darkness, as when Grimm's men "moved in a grave and slightly awe-inspiring reflected light which was almost as palpable as the khaki would have been which Grimm wished them to wear" (343). Finally, and perhaps most interestingly, it can create imaginary light, as when light is "perhaps," "maybe," or "probably" evocative of something.

It is part of the novel's luminous poetics that descriptions of the immersion into a specific milieu are described as immersions into a specific light. At some point during his journey, Christmas enters various media into which he submerges himself and with which he interacts. When he sits reading in an idyllic spot of nature, for instance, Faulkner describes a moment of a truly reciprocal "coupling" of man and environment. It is not only the human being who observes the environment, the environment also observes him. The relation between the landscape and himself is carried by a comparison he makes between the "yellow day" and a cat: "The yellow day opening peacefully on before him, like a corridor, an arras, into a still chiaroscuro without urgency. It seemed to him that as he sat there the yellow day contemplated him drowsily, like a prone and somnolent yellow cat" (85); carried by "the single trivial combination of letters in quiet and sunny space" (86).

Later, in the evening—Faulkner times it at "nine o'clock" (86)—Christmas enters the milieu of Freedman Town, where first a "curve" of voices awaits him. Descending into the town, he is once more entering the "element of race,"

> surrounded by the summer smell and the summer voices of invisible negroes. They seemed to enclose him like bodiless voices murmuring, talking, laughing, in a language not his. As from the bottom of a thick black pit he saw himself enclosed by cabin shapes, vague, kerosene lit, so that the street lamps themselves seemed to be further spaced, as if the black life, the black breathing had compounded the substance of breath so that not only voices but moving bodies and light itself must become fluid and accrete slowly from partical to particle, of and with the now ponderable night inseparable and one (87–8).

In the scene, elements of light and of space intertwine. Bodies take on a luminous quality although it is almost dark. Like light, "streets radiate." Lips "glare." The town lies trembling *in* the lamplight like a fata morgana:

> Then he could see the town, the glare, the individual lights where streets
> radiated from the square. … And further away and at right angles, the
> far bright rampart of the town itself, and in the angle between the black
> pit from which he had fled with drumming heart and glaring lips. No
> light came from it, from here no breath, no odour. It just lay there,
> black, impenetrable, in its garland of Augusttremulous light. (89)

This is a bit like the poetics of light that Faulkner "takes" from the movies, most directly, perhaps, in *Sanctuary*; noir descriptions that single out visual details, as when "his face, his chin where the light fell on it, was still" (161), when "he found her in the dark exactly where the light had lost her" (177) or when from far way, Grimm, whose "face had that serene, unearthly luminousness of angles in church windows" (347) sees Christmas, "as if" he were photographed: "He saw the fugitive's hands glint once like the flash of a heliograph as the sun struck the handcuffs" (347). Other moments of a "noir light" define, symptomatically, fights, such as the one between Christmas and Brown: "The blow cut his voice short off; moving, springing backward, he vanished from the fall of light, into the darkness" (206), or when he has hijacked the car with the young, terrified couple in it: "He saw only the two young, rigidly forward-looking heads against the light glare, into which the ribbon of the road rushed swaying and fleeing" (214).

The description of an "Augusttremulous light" brings me back to the novel's title, which many critics have thought derived from a colloquial use of the word "light" as meaning to give birth—as when a cow will give birth and be "light" again—and have connected this to Lena's pregnancy. In this case, the title would in fact have nothing to do with light at all, and designate a mere date: In August she, perhaps Lena, was light again; in the sense of no longer heavy. As Ruppersburg notes, however, Faulkner changed the title after a casual remark by his wife Estelle, who asked Faulkner: "Does it ever seem to you that the light in August is different from any other time of the year?" (Ruppersburg 1994: 3). Hearing the sentence, Faulkner jumped up, said, "That's it!" and changed the title from *Dark House* to *Light in August*. As Ruppersburg goes on to quote Faulkner:

> In August in Mississippi there's a few days somewhere about the middle of the month when suddenly there's a foretaste of fall, it's cool, there's a lambence, a soft, a luminous quality to the light, as though it came not from just today but from back in the old classic times. It might have fauns and satyrs and the gods and—from Greece, from Olympus in it somewhere. It lasts just for a day or two, then it's gone ... the title reminded me of that time, of a luminosity older than our Christian civilization. (3)

The obsession of Faulkner and his characters with the past is quite literally carried by the light. A classic milieu in which the Christian South becomes, perhaps, one of the mythical landscapes in Poussin or Lorrain. In the luminosity and sonority of his prose, Faulkner writes the myth of the local itself. The myth of the South. His "luminous classicism" is the reason for the universality of Faulkner's work. Like myth itself, his writing is both utterly local and utterly universal.

In his essay "Dante Now: The Gossip of Eternity," George Steiner notes that Dante's *Divina Comedia* is "timeless, universal, because utterly dated and placed" (Steiner 1972: 172). For Steiner, it is Dante and Marcel Proust who "like no others, give us the gossip of eternity" (172). One is tempted to add Faulkner to Steiner's list of luminaries.

# References

Deleuze, Gilles and Claire Parnet (1996), *L'Abécédaire*, dir. Pierre-André Boutang, Paris: Éditions Montparnasse.

Deleuze, Gilles (1986), *Cinema I. The Movement-Image*, trans. Hugh Tomlinson and Barbara Habberjam, Minneapolis: University of Minnesota Press.

Deleuze, Gilles (2006a), *Foucault*, trans. S. Hand, Minneapolis: University of Minnesota Press.

Deleuze, Gilles (2006b), "Michel Foucault's Main Concepts," in *Two Regimes of Madness: Texts and Interviews 1975-1995*, ed. David Lapoujade and trans. Ames Hodges and Mike Taormina, New York: Semiotext(e).

Deleuze, Gilles (2006c), "The Brain is the Screen," in *Two Regimes of Madness. Texts and Interviews 1975-1995*, ed. David Lapoujade and trans. Ames Hodges and Mike Taormina, New York: Semiotext(e).

Faulkner, William (1981), *Light in August*, Penguin: Harmondsworth.

Faulkner, William (1982), *Absalom, Absalom*, Penguin: Harmondsworth.

Faulkner, William (2011), *Sanctuary*, Vintage: London.

Faulkner, William, "Interview": http://faulkner.lib.virginia.edu/display/wfaudio06_1 (accessed February 20, 2013).

Heider, Fritz (2004), "Ding und Medium," in Claus Pias, Joseph Vogl, Lorenz Engell, Oliver Fahle, and Britta Neitzel (eds), *Kursbuch Medienkultur. Die maßgeblichen Theorien von Brecht bis Baudrillard*, Stuttgart: Deutsche Verlags-Anstalt DVA.

Maturana, Humberto (1994), *Was ist Erkennen?* München: Piper.

Maturana, Humberto and Francisco Varela (1998), *The Tree of Knowledge: The Biological Roots of Human Understanding*, trans. Robert Paolucci, Boston: Shambhala.

Maturana, Humberto, "Autopoiesis, Structural Coupling and Cognition: A History of these and other notions in the Biology of Cognition". In: *Cybernetics & Human Knowing*, 9 (3–4) (2002): 5–34, 17.

Ruppersburg, Hugh (1994), *Reading Faulkner: Light in August*, Jackson: University Press of Mississippi.

Schöne, Wolfgang (1977), "Über das Licht in der Malerei". Berlin: Mann.

Steiner, George (1972), "Dante Now: The Gossip of Eternity," in *On Difficulty and Other Essays*, Oxford: Oxford University Press.

Part Three

# [Film|Matter]

# 5

# Figure|Ground: Stills from the Films of Bill Morrison

## Bill Morrison

[all images courtesy of bill morrison–hypnotic pictures]

Grandmother (from "Spark of Being")

Grandson (from "Spark of Being")

**Figure 5.1**

from "Light Is Calling"

**Figure 5.2**

Hitchhikers (from "Spark of Being")

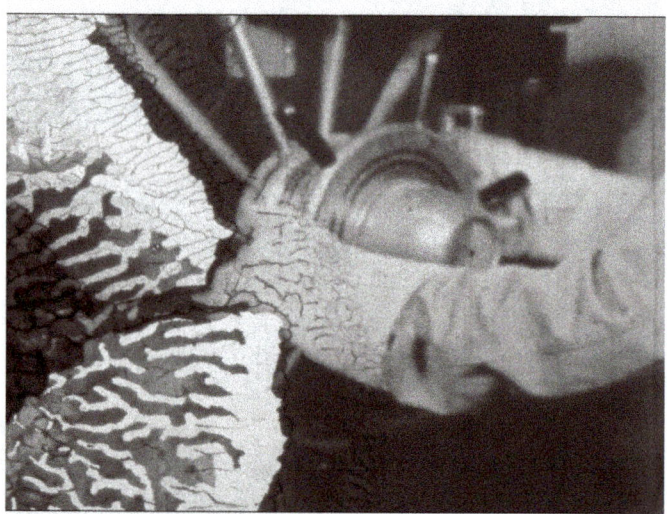

Vault (from "Spark of Being")

**Figure 5.3**

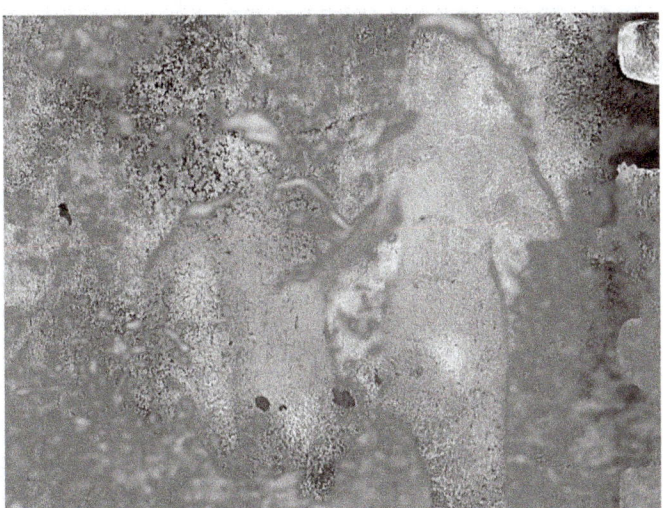

Running Couple (from "Spark of Being")

**Figure 5.4**

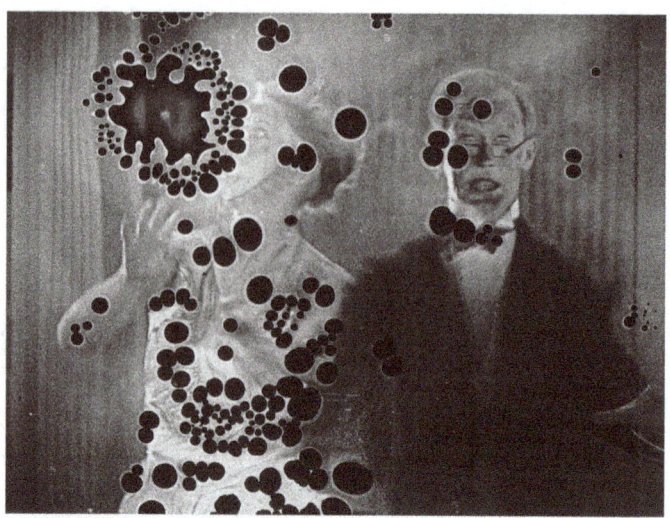

Dancers (from "Decasia")

Laughing Woman (from "Decasia")

**Figure 5.5**

Judge (from "Decasia")

Old Woman (from "Decasia")

**Figure 5.6**

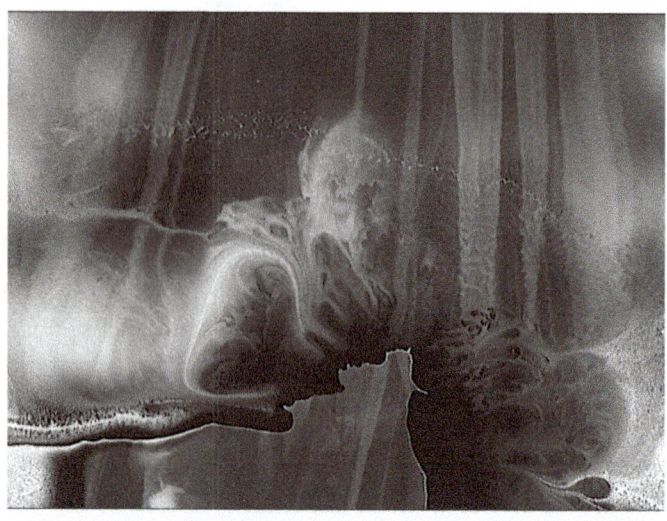

from "All Vows"

**Figure 5.7**

from "Beyond Zero: 1914-1918"

**Figure 5.8**

The Ascension (from "Just Ancient Loops")

**Figure 5.9**

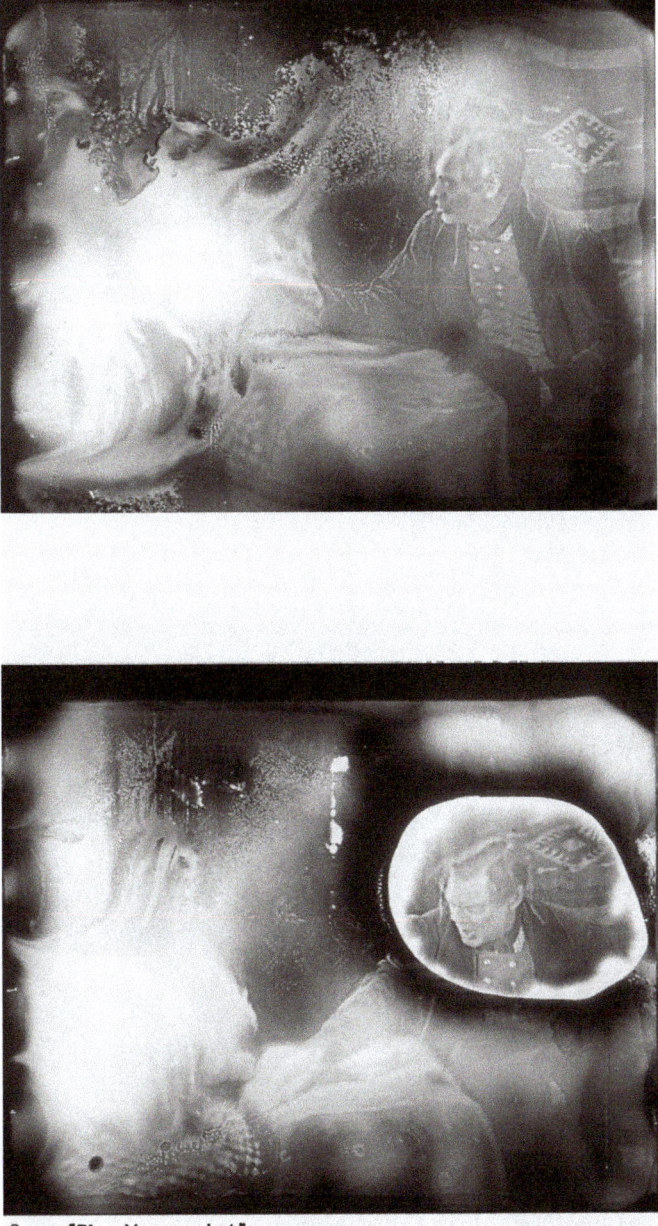

from "The Mesmerist"

**Figure 5.10**

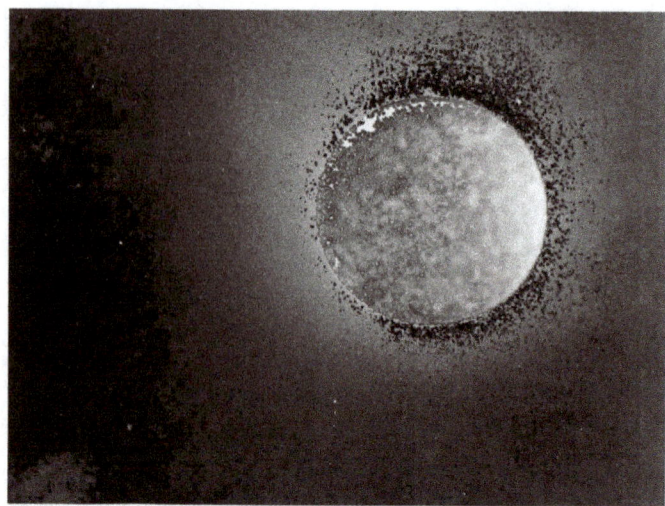

Moon (from "Just Ancient Loops")

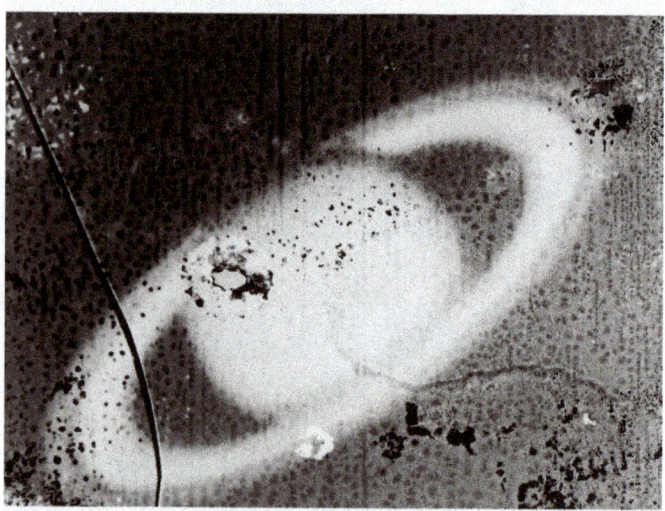

Saturn (from "Just Ancient Loops")

**Figure 5.11**

# 6

# Matter that Images:
# Bill Morrison's *Decasia*

Bernd Herzogenrath

The year 1995 was an important year for film and media studies in at least two respects. The year when "the cinema" celebrated its 100th anniversary, Sony, Philips, Toshiba, and Time Warner agreed on a standard for a data carrier, formerly known as Digital *Video* Disk—the DVD (Digital Versatile Disk), that on the one hand declared war on "the cinema as we know it," but on the other hand promised salvation: the medium film, having since its early beginnings sworn to "capture" movement and the dynamics of life, had to struggle against its transience more than any other medium. In the year of its 100th anniversary, the cinema was not only "old," an "old-fashioned-next-to-outdated" medium; the films themselves, the collected and archived reliquaries of film history, were in danger of rotting, decaying, and disappearing forever. Judging from the password of film conservationists—"From the conservation of the medium to the preservation of the content" (Schüller)—the DVD (or, in general: digital media) in fact seemed to be the redeemer that "film" had longed for. This force field of the hope of "making the moment stay forever" and the dread of decay, this oscillation of materiality and immateriality, of the animation of the static and the reanimation of *le temps perdu* reenacts 100 years later the relation of film, time, life, and death that already had marked the first steps of the medium film—history repeats.

The year 1995 also was the year in which the *Journal of Material Culture* was conceptualized, in order to give a public and interdisciplinary face to a field of research that had already begun to take hold in various disciplines such as anthropology, archaeology, and geography. During the past thirteen years, material culture studies advanced to a new, exciting, and highly influential field of cultural studies.[1] Material culture is based on the premise that the *materiality* of objects is an integrative part and parcel of culture, that the material dimension is as fundamentally important in the understanding of a culture as language or social relations—material culture thus adds a welcome counterweight and addition to the domination of cultural studies by social|linguistic constructivism. Materiality has significance independent of human action or intervention—it is as important to ask how things *do* things (and what kind of things things do), as it is how to do things with words. Objects have a life of their own, a temporality of their own, "objects change over time, in both their physical composition and their cultural salience" (Eastop 2006: 516).

Since material culture studies mainly focuses on the materiality of everyday objects and their *representation* in the media (literature, film, arts, etc.), a further and important step would be to redirect such an analysis to the materiality of the media *itself*, to point the probing finger not only at the thing *in* representation, but at the thing *of* representation. The medium "film" seems to me most fitting to test such an interface of material culture and media studies, since film has entertained a most complex relation to *time* from its early beginnings onward: film promised to (re)present temporal dynamics—and the temporality of things—*directly*, un*media*ted, a paradox that gives rise to the different "strategies" of what Deleuze calls the *movement-image* and the *time-image*, respectively. Such a representation, however, is an effect not only of a perceptive illusion, but also of the *repression* of the very materiality of film itself, the film stock, an immensely fragile medium that in the course of its "projection-life" is subjected to scratches, burns, etc.—to signs of the times. I will situate this crossbreed of material culture and media studies in the larger framework of Deleuze's *Cinema* books mixed with his "intelligent materialism"[2]—a hybrid that stays in the family, so to speak, in order, as Régis Debray put it, "to proceed as if mediology could become in relation to semiology what ecology is to the biosphere. Cannot a 'mediasphere' be treated like an ecosystem, formed on the one hand by populations of signs and on the other by a network of vectors and material bases for the signs?" (Debray 1996: 108).[3]

The following essay focuses on this nexus of film, time, and materiality. I will begin by introducing film's constitutive|constituting move as the attempt to *represent* time *in* film, which was already being discussed at the birth of the medium. Taking my cue from Bazin's influential article on "Ontology of the Photographic Image" (a kind of inspiration for Deleuze's own work on film as well), which also tries to answer the question *What is Cinema?*, I will shift my focus to the *materiality* of film: time leaves much more direct traces *on* film than any representation of time *in* film could ever achieve. Taking Bill Morrison's film *Decasia* (2002) as an example, I will then self-reflexively direct the material culture approach to the filmic *material*. If such an interest in the "possibilities" of the celluloid had already driven much of the 1960s "avant-garde" (Brakhage, Jacobs, etc.), *Decasia* in addition does focus not only on film's "thingness," but also on its own, particular "temporality." Put together from *found footage* and archive material in various states of decay, this film reveals the "collaboration" of time and matter as *in itself* "creative," and ultimately produces a category that I will call the *matter-image* and that, I argue, neither Deleuze's *movement-image* nor his *time-image* completely grasps: here, *time and matter produce their own filmic image.*

## Film: Time|Movement

## Film: Time|Movement—*projection*

Since its birth, the cinema has entertained a complex relation with time. First of all, film was seen as a medium of *representing* time. Marey's chronophotography here clearly can be seen as one of the "midwives" of film. By creating ever smaller temporal equidistances in the measuring, fragmentation, and representation of time, Marey wanted to lift the veil of the mystery of "living machines." According to him, chronophotography proved once and for all that "motion was only the relation of time to space" (Marey 1885: xi). This puts Marey in direct opposition to Henri Bergson's philosophy of time—Bergson explicitly understood time *not* in its reduction to movement in space. It thus comes as no surprise that Bergson entertained a skeptical or at least ambivalent attitude toward the cinema. In his 1907 study *Creative Evolution*, Bergson reveals what he calls the mechanistic "contrivance of the cinematograph" (Bergson 1944: 332)—it "calculates" movement out of "immobility set beside immobility, even endlessly" (331). If, as Marey had claimed, movement is only "the relation of time to space," then, Bergson argues, "time is made up of distinct parts immediately adjacent to one another. No doubt we still say that they follow one another, but in that case succession is similar to that of the images on a cinematographic film" (Bergson 1992: 18), and this completely misunderstands the fundamental difference between time as becoming, as continuous production of newness in the dynamics of an endless differentiation of life, and time as a "mechanic" succession of moments "cut out" of that very continuum. Bergson's *duree* has to be understood as a heterogeneous, qualitative duration, which is completely at odds with Marey's quantitative, numeric, and linear conception of time as *temps* [t]—an opposition that finds its filmic equivalent in the tension between the single image and the projected film.

## Film: Time|Movement—*representation*

The classic narrative film *represents* time *in* film with well-known narrative strategies such as organic montage, rational cuts, continuity editing, and flashbacks, hence, with the action-reaction model. Even in its connection with more complex *plots* (see *Back to the Future*, or *Memento*), narrative film is ultimately based on the concept of an abstract and linear time—exactly what Marey had in mind.

Films based on the action-reaction schema are films that in the Deleuzian taxonomy belong to the *movement-image*. Deleuze argues that when the reality of the Second World War and its aftermath exceeded our capacity for understanding, traditional forms of cinematic "cause-and-effect" strategies became irritatingly inappropriate, resulting in the "crisis of the action-image" (Deleuze 1986: *Cinema 1*, 197) and the breakdown of its corresponding "realist fundament," the "sensory-motor schema" (155). Here, continuity was basically the effect of the filmic characters' movement through space—rational intervals ensure continuity, and the actors function as differentials to translate dramatic action into movement, propelling a cohesive narrative forward.

Through this pragmatic arrangement of space, the organic regime of classic cinema established a spatial continuity based on the movement of its protagonists. Action extends through rational intervals established by continuity editing so that the actor's translation of dramatic action into movement provides the primary vehicle by which a cohesive narrative space unfolds. Since the war, as Deleuze points out, dramatically "increased the situations which we no longer knew how to react to, in spaces which we no longer know how to describe" (Deleuze 1986: *Cinema 2*, xi), the "action-image of the old cinema" (xi) fell into crisis.

As a result, the rational cuts and the continuity of the sensory-motor linkage loosen and collapse—the emerging interval marks the convergence of discontinuous durations and gives way to "false continuity and irrational cuts" (xi). In postwar's "any spaces whatever" (xi), the deserted *Trümmer*-wastelands of, for example, Italian neorealism, movement comes to take on "*false*" (xi) forms, which delink and uncouple continuity, allowing "time 'in its pure state' [to rise] up to the surface of the screen" (xi). The resulting time-image emerges as something *beyond* movement (see 1–24), an image not defined as a succession of spatial segments, subverting the sensory-motor schema and not treating time as a simple derivative of space. According to D. N. Rodowick, "the founding question" of this second regime is, "how to distinguish movement in time from movement in space" (Rodowick 1997: 79). No longer a measure of objects changing their positions *in space*, movement becomes a dynamics of relations *within time*.

## Film: Time|Movement—*preservation*

A further, no less important relation between film and time lies in film's attempt to *preserve* time, in its promise to not only *represent* time, but also actually *capture* and *freeze* it in its fleeting dynamics. After the first screening of Lumière's *actualities* at the Salon Indien in Paris, on December 28, 1895, the daily newspapers celebrated the "fact" that this new medium, with its possibility to record people "in life," made death lose its sting: "We already can collect and reproduce words; now we can collect and reproduce life. We might even, for instance, see those as if living again long after they have been gone" (*Le Radical* December 30, 1895)—"When apparatuses like this are available to the public, when everyone can photograph those that are dear to them, not only their posed forms, but their movements, their actions, their familiar gestures, with words at the tips of their tongues, death will cease to be absolute" (*La Poste* December 30, 1895).[4] Death is also the central term in André Bazin's discussion of photography and film in his influential essay, "The Ontology of the Photographic Image." Bazin here claims an anthropological cause for the arts in general, which he calls a "mummy complex" (Bazin 2005: "The Ontology of the Photographic Image," 9). Like the "practice of embalming the dead" (9), which aimed at the "continued existence of the corporeal body" (9), the image was to provide an almost magical *and* material "defense against the passage of time" (9), with the aim of "the preservation of life by a representation of life" (10). For Bazin, "death is but the victory of time" (9). Similarly, as follows from Bazin's "integral realism,"[5] photography and film are the victory over time, over forgetting, the "second spiritual death" (10), conserving time "by means of the form that endures" (10). Art as a means to immortalize man—Bazin is catching up with a traditional *topos* here. But in contrast to traditional painting's "obsession with likeness" (12)— C. S. Peirce would call this "iconological character"—photography rather is a "molding, the taking of an impression, by the manipulation of light" (12*n*†), an index, a "tracing" (Bazin 2005: "Theater and Cinema," 96) of a human being or an object. Thus, photography mummifies *the* *moment* in its "transference of reality from the thing to its reproduction" ("Ontology," 14), but this mummification, due to its very instantaneity, is compelled to "capture time only piecemeal" ("Theater and Cinema," 96). Still—photography shares with film the "indexical character"—film, like photography, is "the art of the index; it is an attempt to make art out of a footprint" (Manovich 2008). However, a film is marked by a surplus advantage—"it makes a molding of the object as it exists in time and, furthermore, makes an imprint of the duration of the object" ("Theater and Cinema," 97; see also Deleuze, *Cinema 1*, 24). The mummy *of* film (like the mummies *in* film) lives (as every film-lover knows, and Bazin knew as well)!! Bazin's mummy has a twofold function—it conserves the recorded image and it dynamizes the otherwise static image. By means of the filmic mummy, as Bazin famously put it, "the image of things is likewise the image of their duration, change mummified, as it were" ("Ontology," 15). In the only illustration to Bazin's "Ontology," we get an image of the Holy Shroud of Turin (Figure 6.1), which is defined by Bazin as a synthesis of "relic and photograph" (14*n*\*). This allows us, I argue, to deduce that Bazin in analogy sees the filmic material, the actual celluloid carrier, as the mummy's shroud or bandage, and the balm or preserving natron as a kind of emulsion that makes possible a direct "fingerprint" of the real, so that precisely photography's|film's "automatism" devoid of an intervening subject (which coincides with Bazin's idea of realism) makes "the logical distinction between what is imaginary and what is real ... disappear" (15). As already mentioned, film "embalms" time, "rescuing it simply from its proper corruption" (14). But what if the corruption and entropy proper to time also eat at the mummy's bandages? What if these die and decay, which also means—what if these have *a proper life of their own*?

## Film: Time|Movement—*manifestation*

This Film is Dangerous![6] I am not referring to the contents of movies that supposedly are corrupting our youths, films containing "scenes of

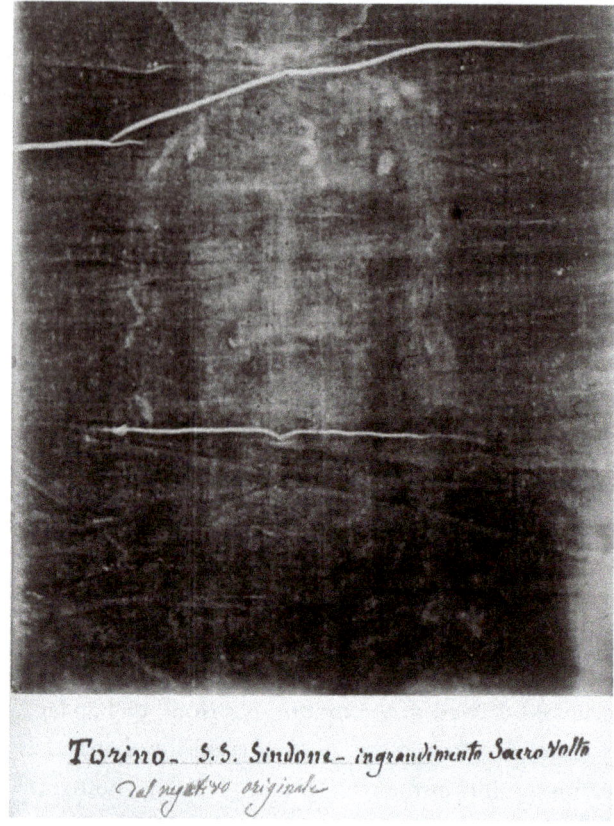

*Torino_ S.S. Sindone_ ingrandimento Sacro Volto*
*dal negativo originale*

**Figure 6.1** The holy shroud of turin—a synthesis of "relic and photograph" (Wiki Commons).

nudity and extreme violence"—I want to focus on the *material* level of film, not on the level of narration, nor of technology and techniques, but on the fundamental level of the film's *thingness*—the film strip, a.k.a. "celluloid." Until approximately 1950, all movies were shot on nitrate film, on nitrocellulose (commonly referred to as "celluloid"), a highly inflammable material—just remember the scene in Giuseppe Tornatore's *Cinema Paradiso* (or Tarantino's *Inglorious Basterds*, with its "Operation Kino"), where the cinema goes up in flames. Developed in 1899 by George Eastman, the immense advantage of nitrate film was its high quality—no other material provided such brilliance and high amount of shades of gray. But nitrocellulose consists of cotton, camphor, and acid and is based on the same formula as the so-called gun cotton—nitrate film carries loads of oxygen in its own pockets to fend the flames, so that it even burns under water (Figure 6.2).

In addition, once processed, this material is highly sensitive to "environmental factors": it tends to decompose and deteriorate in dependence of time and environment, and it returns to its components—nitrocellulose, gelatin, and silver emulsion. This process enfolds in various states; it begins with a sepia/amber "coloration" of the film strip and the fading of filmed images; then the celluloid loses its "shape," softens, and becomes gooey; in a next step, bubbles and blisters emerge on the surface of the film, the emulsion separating from the nitrocellulose carrier. In the end, the nitrocellulose base completely depolymerizes and hardens into the notorious "hockey pucks" and "donuts" so dreaded by film archivists, until what is left is just a highly inflammable reddish powder.

Bazin saw the medium of film as a bandage, as a protective skin—in French, the material film strip is referred to as *pellicule* (skin). Since the film (and the skin of film)[7] is also a *thing*, a material object, it is *itself* subjected to time—and to decay—as well. If an actor|actress reaches

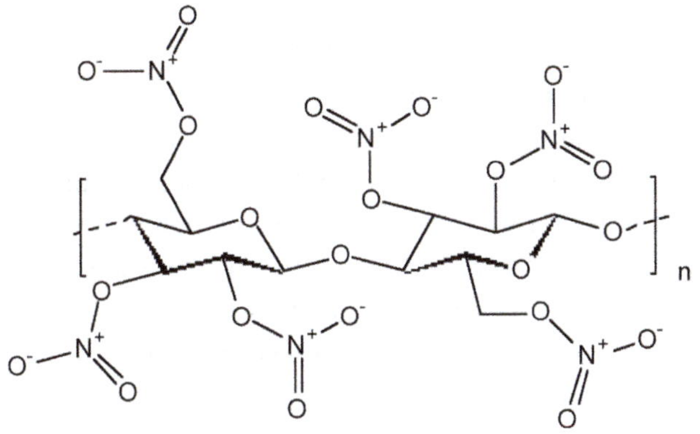

**Figure 6.2** Nitrocellulose formula (Wiki Commons).

an age when s|he loses attraction with the audience, s|he either has a "skin job" or quits acting. Likewise films, if time has left too many marks on their surface, are being restored ("embalmed") or taken out of distribution. The entropic process can be slowed down, but it cannot be stopped—and it is exactly these decaying film skins that Bill Morrison uses as basic material for his film *Decasia* (2002). *Decasia* takes film's materiality seriously and lends itself to a "materialist approach" to media studies—representation of time and things *in* film are complemented by a perspective that takes into consideration the temporality of the medium *itself.*

## *Decasia*: The matter-image

## Film is also a thing ...

Morrison's *Decasia* can be located in the tradition of the American avant-garde or experimental film of the 1960s and 1970s. A main characteristic of this tradition was its focus on the filmic material and on the structure of film, and not so much on content and narration. Filmmakers such as Bruce Connor, Robert Breer, and Tony Conrad worked with the concept of *flicker*-film that undermined classic filmic temporality (and its concomitant continuity-effect)—twenty-four frames per second—and experimented with various tempi. Andy Warhol rediscovered early cinema's stylistic device of the "static camera" and made *duration* the explicit topic of films such as *Empire, Sleep,* and *Eat.* Ken Jacobs, George Landow, and others utilized the concept of *found footage* for the experimental film, while Stan Brakhage produced films completely without a camera, by what Peirce would have called "indexical" procedures—putting objects directly on the film strip to be processed, painting or scratching on its surface, etc. (see, e.g., Deleuze 1986: *Cinema 1,* 84–5). It was Brakhage's self-expressed aim to decouple the filmic image from its hegemonic relation to memory, to deconstruct the images' representational character, and to create a "sense of constant present-tense" (210)—not a representation of the past, but a presentation of temporalities, of durations. As P. Adams Sitney has put it, American experimental film of the 1960s and 1970s were facing "the great challenge ... of ... how to orchestrate duration" (351–2).[8] Common to all these experiments was the desire to make the filmic material *itself*—under "classic circumstances" invisible due to the ideal of the transparency of the medium according to which film is "the material base that must be dematerialized in projection" (Stewart 1999: 3)—visible and fruitful as a fundamental component of the filmic process.

Morrison goes a decisive step further—*Decasia* is a montage made from found footage films in various states of decay. He leaves the sequences basically untreated in order to present a *time-image* created not by a human subject, but by time and matter itself—the *matter-image*. In order to get his material, Morrison had been digging his way through various film archives—like Walter Benjamin's "rag picker" (*Lumpensammler*)—Morrison searched the archives of the Library of Congress, and the archive of 20th Century Fox Movietone Newsreels at the University of North Carolina, in particular their collections of *actualitées*, travel reports, industrial and educational films that all dated from the first half of the Century of the Cinema and that all were shot on nitrate film.[9] In a way, I argue, Morrison's strategy enacts a reversal of classic cinema's subordination of time to movement comparable to the Deleuzian taxonomy. *Decasia's* cannibalization and recontextualization of prewar "movement-images" according to irrational cuts and false continuities enacts an undermining of the concept of time as the relation of movement and space. Whereas in the classic movement-image, the rational cut served as a "linkage of images" (Deleuze 1986: *Cinema 2*, 213), producing "natural relations (series)" (Deleuze 1986: *Cinema 1*, 204) of images, the film of the time-image "'disenchain[s]'" (Bogue 2003: 173) the images from these series, opening up and expanding an "irrational interval" by which each image, according to Rodowick,

becomes what probability physics calls a "bifurcation point," where it is impossible to know or predict in advance which direction change will take. The chronological time of the movement-image fragments into an image of uncertain becoming ... the regime of the time-image replaces this deterministic universe with a probabilistic one. (Rodowick 1997: 15)

This is not to suggest that *Decasia* is a random collection of images and sequences—quite the contrary, in an interview Morrison reveals his thorough composition of the film.[10] However, the relation between images and sequences is undetermined, unpredictable, and probabilistic.

*Decasia* begins (and ends) with the image of a spinning Sufi dancer from Egypt (Figure 6.3)—Bazin's country of mummies. Already at the beginning, *Decasia* accentuates the paradox of what could be called a "static dynamics"—here, movement does not propel a plot by action-reaction, but rather stays "within the frame," and within the confinements of this frame, movement "happens" only locally, as if space does not exist (or matter), whereas the movement itself deconstructs its proper "motor function" and allows a glimpse of what Deleuze calls "a little time in the pure state" (Deleuze 1986: *Cinema 2*, xi). Thus, as

**Figure 6.3** Sufi dancer in *Decasia* (courtesy of Bill Morrison—Hypnotic Pictures).

Rodowick explains, "to the extent that time is no longer the measure of movement as indirect image, movement becomes a perspective on time" (81), a *direct* time-image, independent of montage strategies. After the Sufi dancer, a sequence shot in a film laboratory and rotating film reels follows—the audience witnesses the birth of a film in film.[11]

The dancer's circular movement is taken up again in this sequence and enacts the constituting paradox of the filmic medium: the "static dynamics" *of* film—movement and stasis *at the same time*, the illusion of movement as the effect of static snapshots is complemented by the "static dynamics" *in* film (the Sufi dancer), subverting or at least questioning the sensory-motor schema of the classic movement-image. Images of movement and circulation, of birth, life, and death, provide a "red thread" in Morrison's film and are also taken up in the circular structure of *Decasia* itself, opening and concluding with the Sufi dancer. "Repetition" is one of Morrison's stylistic means—he often uses the same "parent movie" (*found footage*) in various films. In *Decasia*, he uses sequences already used in his earlier films, such as *The Film of Her* (1997) and *Trinity* (2002). However, re-petition—just like re-membrance—is not a repetition of the same ... this would rather be re-dundancy. Morrison rather "extracts" sequences from their "original" narratives and embeds them in a new context—in the context of time itself. The "return" of certain images returns as difference, and thus has a certain affinity to memory, as Morrison himself points out:

> The frame pauses briefly before the projector's lamp, and then moves on. Our lives are accumulations of ephemeral images and moments that our consciousness constructs into a reality. No sooner have we grasped the present, it is relegated to the past, where it only exists in the subjective history of each individual. (2006)

After the two intro-sequences, scenes and images in various states of decomposition and decay follow. *Decasia* does not see the signs of the time as flaws, as material defects—they rather transfer their own aesthetics onto the images. Morrison has deliberately chosen sequences where the representation engages in direct contact with the material carrier. A boxer is seen fighting against an amorphous blob (once presumably the image of a punching ball) threatening to swallow him (Figure 6.4).

"Flames" are dancing over the close-up face of a woman, "wounding" both celluloid and image. The film's|woman's skin cracks and bubbles and seethes like molten lava—the woman's face gets "out of shape," melts. The subject|title of the film seems to have transferred|inscribed itself into its material. The resulting tensions create a texture so porous it recalls "a 'pointilliste' texture in the manner of Seurat" (Deleuze 1986: *Cinema 1*, 85),[12] and produce cracks that echo old oil paintings, but also of some of Brakhage's works. *Decasia* owns a tactile texture, an almost sculptural depth missing from most contemporary film—this is not the utopia of the digital image, as sharply defined as possible, but the idea of an almost three-dimensional geology of surface. Morrison's approach starts with the materiality of the filmic medium and its own *proper* metamorphosis, rather than its capability to represent time and things—the temporality and thingness of the material *itself* is the center of his work, not the forms and shapes it *represents*, but the shape and form it *becomes*. The struggle between

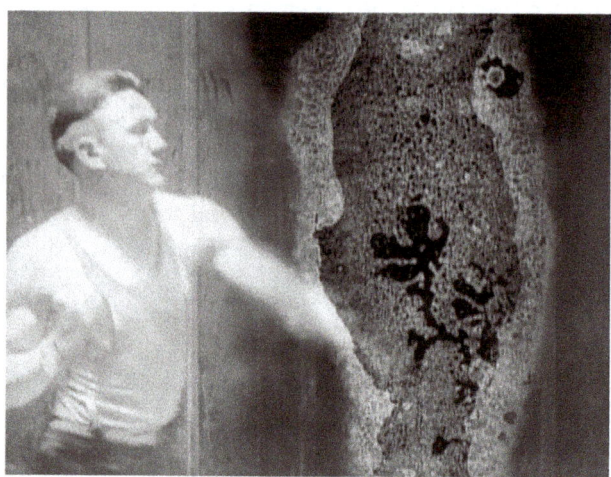

**Figure 6.4** Boxer in *Decasia* (courtesy of Bill Morrison—Hypnotic Pictures).

image and material *ruins* the narration of the "original film," but produces a new "narrative" that *Decasia* does not *illustrate*, but that emerges out of the ruinous image *itself*.

The return of film's (repressed) materiality makes itself seen as the destruction of the image which it had produced in the first place—yet, as Joachim Paech has poignantly stated, "the death of images ... is itself an image again, otherwise it would not be representable" (123).[13] In Morrison's matter-image, film is revealed as *image-producing materiality*, not as an illusion of reality, as in classic film. Since, for the audience of *Decasia*, the (re-)entry of the material in the medial form appears as the very destruction of that form, the result is a paradoxical *mise-en-scène* of the simultaneity of appearance and disappearance, of destruction and construction. The filmic material is not (only) a *transparent transmitter* of images and meaning, but rather *instrumental* in its construction—the subject of "time" in *Decasia* is *presented* on the filmic material *directly*, by the material's "treatment" *by time itself*.

## Ruinous film|Filmic ruin

Morrison's films constitute and partake what might be called a "poetics of the ruin,"[14] a poetics of the historicity of film not in the sense of traditional historiography of film, but with regard to the historicity— even "mortality"—of its thingness. From this perspective, film history becomes the history of film's decay, which, according to Paolo Cherchi Usai, makes a history of film possible in the first place: "Such images [that are immune from decay] can have no history" (Usai 2001: 41). Everything "happening" to a film from its "birth" to its "death" constitutes its history—if all films would be unharmed by time and "survive," there would be no history of film—"cinema is the art of destroying moving images" (7). However, *Decasia* does not really fit into the tradition of "images of ruins" of (post) 09|11 cinema—*Decasia* rather presents "ruinous images"; it is a "ruinous film|filmic ruin" that does not *represent* the decay of *some other object*, but *enacts* the decomposition *of its own material*.

These ruinous images deconstruct the linear time of classic film—they seem to emerge from the fringes of "readability," located between pure indexicality and meaning, between a "re-animated present of the past" and time as a complex mystery. Film's mythical power to "capture time" merges with the tragedy that the medium film *itself*—as materiality—is also subjected to the vicissitudes of time—here, the poetics of the archive[15] is married to the poetics of the ruin; indexicality connects with entropy.

Here, film leaps over the threshold separating the "likeness-factor" of representation from direct "embodiment"—C. S. Peirce has theorized this in semiotic terms as the difference between *icon* and *index* and has pointed out that, for example, in photography, the iconic relation of likeness is only a secondary and forced effect of its indexicality:

> Photographs, especially instantaneous photographs, are very instructive, because we know that they are in certain respects exactly like the objects they represent. But this resemblance is due to the photographs having been produced under such circumstances that they were physically *forced* to correspond point by point to nature. (5–6, emphasis added)

In *Decasia*'s "ruinous images," the indexicality is not only the one underlying the iconicity of the represented figures and objects, but also first and foremost an index that is a chemical reaction of the compounds of nitrocellulose with the environment. And *Decasia*'s represented figures and forms do not deteriorate because of a diegetically motivated decomposition (as in the Horror Film—see, e.g., the early films of David Cronenberg, or Philip Brody's *Body Melt* (1993)), but because of the decay of its carrier materiality. This logic of matter's "reclaiming of power" against its forced (in)formation by man is the central topic of Georg Simmel's essay "The Ruin" (1907). The "artistic formation" enacted by the creative subject (Simmel refers to architecture in particular) here appears as an "act of violence committed by the spirit to which the stone has unwillingly submitted" (Simmel 1959: 260)—there is a similar "physical force" at work like the one underlying the iconic aspect of the index. In a ruin, however, "decay destroys the unity of form" (260), spirit engages in a dialectical struggle with nature and with the "laws governing the material" (*Eigengesetzlichkeit des Materials*) (259)—and this material aims at putting a stop to the subject's|the spirit's game. From "the standpoint of ... purpose" (260), from the perspective of the "unity of form," this *natural* decay appears as "a meaningless incident" (260)—however, the result of this is not the simple "formlessness of mere matter" (261). The fascination of the ruin—and of a ruinous film such as *Decasia*—is precisely the fact that the destruction of an object (or of an image) makes a new object|image emerge, a "new form which, from the standpoint of nature, is entirely meaningful, comprehensible, differentiated" (261–2). This "new form" is the result of antagonistic forces, of the interplay of entropy and evolution, of the past and present, intention and chance. The ruin—like Morrison's *Decasia*—simultaneously struggles *and* plays with its own destruction, and in this very oscillation a "new form" emerges. Thus, in *Decasia*, scenes in which the amorphous mass threatens to swallow the "diegetic life" are on a par with scenes in which the image precisely seems to emerge out of that blob (Figures 6.5 and 6.6).

All things considered, the ruin lacks nothing—above all, it does not lack any "preceding totality": the ruin does not only provide its own aesthetic criteria (as Ralph Waldo Emerson put it: "Even the corpse has its own beauty" (14)). Strictly speaking, only from a human, "purposive" perspective, one can talk of entropy and decay—the "arrow of time," as Bergson points out, is the necessary condition of the creation of newness:

> The living being essentially has duration; it has duration precisely because it is continuously elaborating what is new and because there is no elaboration without searching, no searching without groping. Time is this very hesitation. ... Suppress the conscious and the living [of the material world] ..., you obtain in fact a universe whose successive states are in theory calculable in advance, like the images placed side by

**Figures 6.5 and 6.6** Two images from *Decasia* (courtesy of Bill Morrison—Hypnotic Pictures).

side along the cinematographic film, prior to its unrolling. ... Would not the existence of time prove that there is indetermination in things? Would not time be that indetermination itself? (Bergson 1992: 93)

In its continuous folding of past into present and vice versa, with the ruin (as with *Decasia*, with its similar folding of outside (materiality) into inside (image) and vice versa) one cannot simply designate "decay" as the negative, destructive force: like with the Moebius Strip, the outside is simultaneously part of the inside, decay and composition become indiscernible, being destructive and creative *at the same time*. If in the abstract *temps* of Marey (and of Classic Physics and of Classic Film), as Bergson maintains, there can be no creation, and if this statement remains true for the "narrative level" of film, on the level of the materiality of the medium, *newness emerges*.

## The aura of the thing

When Simmel describes the patina on metal, wood, ivory, and marble, it almost seems as if he was talking about the images in *Decasia* and the "mysterious harmony" that "the product becomes more beautiful by chemical and physical means; that what has been willed becomes, without intention of force, something obviously new, often more beautiful" (Simmel 1959: 262), resulting in a "special something" that "no new fabric can imitate" (264).[16] This singularity comes close to what Walter Benjamin has famously designated as *aura*, the work of art's "presence in time and space, its unique existence" (220), which has declined in the age of mechanical reproduction.[17] "Aura" comes close to being the historicity of materiality. According to Benjamin, aura's "analogue in the case of a utilitarian object is the experience that precipitates on this object" ("On Some Motifs in Baudelaire," 188)[18]—the aura of a work of art is a direct effect of its "contact" with time and space. Morrison points out the importance of this "direct contact" as well:

> Older archival footage … [has] this quality of having been touched …
> by time, by a non-human intervention that is organic … there are
> many things happening between the first time they were registered on
> the 35 mm negative and transferred to a paper intermediary, to being
> stored, rained on, or being nibbled by rats; the hairs in the specs, the
> grain and what would have to happen for that to be brought out and to

be re-photographed some sixty years later. So each picture has its own
dimension of time, its own history. Whether or not you are conscious
of this while watching, you are still watching these tiny histories go
by ... (qtd. in Habib "Cinema from the Ruins")

With Morrison "staging" the film as a singular, material object, and
with the continuous oscillation of materiality, filmed objects, and time,
*Decasia* succeeds, I argue, in the "re-auratization" of film precisely in the
age of mechanical reproduction. When Bazin claims that photography
(and implicitly: film) "affects us like a phenomenon in nature, like
a flower or a snowflake whose vegetable or earthly origins are an
inseparable part of their beauty" (Bazin 2005: "Ontology" 13), we can
specify with *Decasia*, that film can affect us as a "natural" phenomenon,
because in an important aspect it *is* a natural phenomenon.

   *Decasia* follows a conception of "cinematic time" different from
that which Bergson saw as the biggest drawback of the cinema—its
fundamental linearity and abstractness. *Decasia*'s time is neither the
duration of the projected film, nor the one of the film's narrative, neither
narration time, nor narrated time, but the time of its material. *Decasia*
contradicts Bergson's claim that cinema can only endlessly repeat "the
same"—*Decasia* rather is the cinematic proof for Bergson's observation
that "*wherever anything lives, there is, open somewhere, a register in which
time is being inscribed ... duration*, acting and irreversible" (Bergson
1944: 20).[19] We are presented a film that merges the "non-subjective"
perception of the camera-eye with the "non-human perception" of matter
itself—in its focus on the "perceptiveness of matter," *Decasia* shows
that film is not only a signifying machine, and|or an image-and-sound
machine, but because of its chemical composition it is also something
like "a chlorophyll- or a photosynthesis-machine" (Deleuze|Guattari
1992: *Anti-Oedipus*, 2). The amorphous shapes of|in *Decasia* result
from the oscillation of the formation|representation of objects, and
the natural and organic processes of the object|matter "film" itself—
representation and materiality, image and thing are being folded into
each other. In a commentary on *Decasia*, Morrison puts this in terms
reminiscent of the terminology of "Embodied Mind Philosophy": "The

images can be thought of as desires or memories: actions that take place in the mind. The filmstock can be thought of as their body, that which enables these events to be seen. Like our own bodies this celluloid is a fragile and ephemeral medium that can deteriorate in countless ways" (2006). In a similar manner, George Lakoff and Mark Johnson argue in *Philosophy in the Flesh*: "What is important is not just that we have bodies and that thought is somehow embodied. What is important is that the very peculiar nature of our bodies shapes our very possibilities for conceptualization and categorization" (Lakoff and Johnson 1999: 19), and it is exactly this, I argue, what *Decasia* shows with regard to the filmic *body*, the *materiality* of the medium "film." *Decasia* is on every level a more complex "history of film," with concepts of "history" and "memory" that goes far beyond the film archivists' idea of the "preservation of contents." Morrison comments,

> I've shown *Decasia* in archival symposiums, and archivists rushed up to me afterwards and were saying: "But you must document what all these are." But … that would defeat the purpose. And it would make it seem a plea for preservation which I'm not actually doing. Certainly none of this work would exist without preservation. I am greatly indebted to them but I'm not saying it is necessarily tragic that time erodes these things because, hey, that's what happens … the magic of cinema is also its fleeting nature, not only its objectual nature. (qtd. in Habib 2004, "Cinema from the Ruins")

As Deleuze, in his reading of Bergson, states, "The past which is preserved takes on all the virtues of beginning and beginning again. It is what holds in its depth or its sides the surge of the new reality, the bursting forth of life" (Deleuze 1986: *Cinema 2*, 92).

*Decasia* takes into consideration that, as Bergson wrote, "memory … is just the intersection of mind and matter" (Bergson 1991: 13). It is this folding of perception into memory and vice versa that defines Deleuze's "crystal of time" (see Deleuze: *Cinema 2*, 68–97)—and in Morrison's *Decasia*, I argue, the "crystallization of time" allows for a very materialist reading.

As Deleuze has beautifully put it: "The brain is the screen," cinema is cerebral, but this screen, this brain, this "cinema isn't theater; rather, it makes bodies out of grains" (Deleuze 2000: 366)—Bill Morrison's *matter-image* does exactly that.

# Notes

1  See, for example, Bill Brown (2003), *A Sense of Things. The Object Matter of American Literature*, Chicago: The University of Chicago Press.

2  "Intelligent materialism" not because it is more intelligent than other "materialisms," but because it grants intelligence and agency to matter itself.

3  Debray further proposes—"We speak about Earth Day. Why not tomorrow, no pleasantry intended, a day London devoted to celebrating celluloid, vellum paper, or vinyl records?" (114).

4  *Le Radical* December 30, 1895: "On recueillait déjà et on reproduisait la parole, on recueille maintenant et on reproduit la vie. On pourra, par exemple, revoir agir les siens longtemps après qu 'on les aura perdus." *La Poste* December 30, 1895: "Lorsque ces appareils seront livrés au public, lorsque tous pourront photographier les êtres qui leur sont chers non plus dans leur forme immobile mais dans leur mouvement, dans leur action, dans leurs gestes familiers, avec la parole au bout des lèvres, la mort cessera d 'être absolue."

5  "... the history of the plastic arts ... [is] ... essentially the story of resemblance, or, if you will, of realism" (10).

6  See Smither.

7  See also Marks.

8  Deleuze also refers to Sitney's analysis in (*Cinema 1*: 86; 230*n*22 and 24).

9  *Decasia* is a collaboration with the American Composer Michael Gordon (one of the founders of the *Bang On a Can* collective) in association with The Ridge Theater, New York—I will not go into the intricacies of that very peculiar image|sound cooperation, since that would be an essay of its own. *Decasia* was conceptualized as a film (like what you get on the DVD), but as a multimedia event, premiered 2001 in Basel, Switzerland, with the Basel Sinfonietta Orchestra, slide projections, a very special stage architecture, etc. Gordon's symphony ventures into the fringes

of sound and works with repetitions, superimpositions, etc.—a sonic
equivalent to Morrison's visual strategies.

10  See Barnaby Welch's interview with Morrison for *High Angle Magazine*
    (2002) on http://www.decasia.com/html/highangle.html, last accessed
    April 14, 2008.

11  Note the parallel to Dziga Vertov's *Man with a Movie Camera* ...

12  Deleuze talks about this texture in the section on "gaseous perception"
    (80–6). While Deleuze here comments on, for example, George
    Landow's and Ken Jacobs' use of decaying found footage, quite
    similar to Morrison's, and the idea of "an image defined by molecular
    parameters" (85), and the impression that here "the film itself seems to
    die" (86, quoting from Sitney), Deleuze, I argue, is more concerned with
    perception and the *projected* image itself rather than its material "coming
    into being." See also (Totaro).

13  My translation of: "das Ende der Bilder ... ist selbst ein Bild, anders wäre
    es nicht darstellbar."

14  See, for example, the essays by André Habib, or the essay by Cadava.

15  See, for example, Cohen, and—of course—Derrida.

16  Nitrocellulose, just like the "old fabrics" that Simmel describes, is
    subjected to "dryness and moisture, heat and cold, outer wear and inner
    disintegration" (264).

17  However—"for the last time the aura emanates from the early
    photographs" ("Work of Art" 224). Benjamin explains this "persistence"
    of aura a.o. with the long exposure times of early photography. In
    the reduction of exposure time—which more than faintly reminds of
    Bergson's *duration*—from various hours to only seconds, Benjamin
    sees an important factor of the decay of aura. Correspondingly, long
    exposure time emerges as a sign of the "technical conditionality of the
    auratic appearance" ("Small History" 248, my translation of "technisches
    Bedingtsein der auratischen Erscheinung"), that "strange weave of space
    and time" (250).

18  My translation of "so entspricht die Aura ... eben der Erfahrung, die
    sich an einem Gegenstand des Gebrauchs als Übung absetzt.," in: Walter
    Benjamin (2002), "Über einige Motive bei Baudelaire" *Medienästhetische
    Schriften*, Frankfurt a/M: Suhrkamp, 32–66, 55. The English
    translation—"[aura's] analogue in the case of a utilitarian object is the
    experience which has left traces of the practiced hand" (188), I argue,
    reduces the object's experience to something done to it by a human

hand, whereas Benjamin leaves that open. Also, the German expression "sich absetzen" also alludes to the chemical process of precipitation, which comes quite handy in my context.

19  When Simmel speaks of "the present form of the past" (266)—a concept in which "the ruin" and "the archive" seem to merge, this simultaneity or coexistence of temporalities is of particular relevance to Bergson, who defines memory as "the prolongation of the past into the present" (*Creative Evolution* 20).

# References

Bazin, André (2005a), "The Ontology of the Photographic Image," in Hugh Gray (ed. and trans.), *What Is Cinema? Volume 1*, 9–16, Berkeley, Los Angeles, and London: University of California Press.

Bazin, André (2005b), "Theater and Cinema," in Hugh Gray (ed. and trans.), *What Is Cinema? Volume 1*, 76–124, Berkeley, Los Angeles, and London: University of California Press.

Benjamin, Walter (1968a), "On Some Motifs in Baudelaire," in Hannah Arendt (ed.), *Illuminations. Essays and Reflections*, trans. H. Zohn, 157–202, New York: Harcourt, Brace & World, Inc.

Benjamin, Walter (1968b), "The Work of Art in the Age of Mechanical Reproduction," in Hannah Arendt (ed.), *Illuminations. Essays and Reflections*, trans. H. Zohn, 217–52, New York: Schocken Books.

Benjamin, Walter (1979a), "A Small History of Photography," in *One-Way Street and Other Writings*, trans. Edmund Jephcott and Kingsley Shorter, 240–57, London, New York: Verso.

Benjamin, Walter (2002b), "Über einige Motive bei Baudelaire," in ed. Detlev Schöttker, *Medienästhetische Schriften*, 32–66, Frankfurt a/M: Suhrkamp.

Bergson, Henri (1944), *Creative Evolution*, trans. Arthur Mitchell, New York: The Modern Library.

Bergson, Henri (1991), *Matter and Memory*, trans. N. M. Paul and W. S. Palmer, New York: Zone Books.

Bergson, Henri (1992), *The Creative Mind. An Introduction to Metaphysics*, New York: Citadel Press.

Bogue, Ronald (2003), *Deleuze on Cinema*, New York and London: Routledge.

Brakhage, Stan (2005), "Inspirations," in Bruce R. McPherson (ed.), *Essential Brakhage. Selected Writings on Filmmaking by Stan Brakhage*, 208–11, New York: McPherson & Company.

Brown, Bill (2003), *A Sense of Things. The Object Matter of American Literature*, Chicago and London: The University of Chicago Press.

Cadava, Eduardo (2001), "'Lapsus Imaginis': The Image in Ruins," *October*, 96 (Spring 2001): 35–60.

Cohen, Emily (2004), "The Orphanista Manifesto: Orphan Films and the Politics of Reproduction," *American Anthropologist*, 106 (4): 719–31.

Debray, Régis (1996), "Toward an Ecology of Cultures," in *Media Manifestos: On the Technological Transmission of Cultural Forms*, trans. by Eric Rauth, 108–32, London: Verso.

Deleuze, Gilles (1986a), *Cinema 1. The Movement-Image*, trans. Hugh Tomlinson and B. Habberjam, Minneapolis: University of Minnesota Press.

Deleuze, Gilles (1986b), *Cinema 2. The Time-Image*, trans. Hugh Tomlinson and R. Galeta, London: Athlone 2000 Press.

Deleuze, Gilles (2000), "The Brain Is the Screen: An Interview with Gilles Deleuze," in Gregory Flaxman (ed.), *The Brain Is the Screen. Deleuze and the Philosophy of Cinema*, 365–73, Minneapolis and London: University of Minnesota Press.

Deleuze Gilles and Félix Guattari (1992), *Anti-Oedipus. Capitalism and Schizophrenia*, trans. Robert Hurley, Mark Seem, and Helen R. Lane, Minneapolis: University of Minnesota Press.

Derrida, Jacques (1997), *Archive Fever. A Freudian Impression*, London and Chicago: University of Chicago Press.

Eastop, Dinah (2006), "Conservation as Material Culture," in Chris Tilley, Webb Keane, Susanne Küchler, Mike Rowlands, and Patricia Spyer (eds), *Handbook of Material Culture*, 516–33, London: SAGE.

Emerson, Ralph Waldo (1983), "Nature," in Joel Porte (ed.), *Ralph Waldo Emerson. Essay and Lectures*, 5–49, New York: Library of America.

Habib, André (2004a), "Cinema from the Ruins of the Archives. Matter and Memory: A Conversation with Bill Morrison," *Horschamp* (November 30, 2004): www.horchamps.qc.ca/new_offscreen/interview_morrison.html (accessed November 14, 2014).

Habib, André (2004b), "Thinking in Ruins. Around the Films of Bill Morrison," *Horschamp* (November 30, 2004): www.horchamps.qc.ca/new_offscreen/cinematic_ruins.html (accessed November 14, 2014).

Habib, André (2006), "Ruin, Archive, and the Time of Cinema: Peter Delpeut's *Lyrical Nitrate*," *SubStance*, 110, 35 (2): 120–39.

Lakoff, George and Mark Johnson (1999), *Philosophy in the Flesh. The Embodied Mind and Its Challenge to Western Thought*, New York: Basic Books.

Manovich, Lev (2008), "What Is Digital Cinema?": http://www.manovich.net/TEXT/digital-cinema.html (accessed November 14, 2014).

Marey, Etienne-Jules (1885), *La méthode graphique dans les sciences expérimentales*, Paris: Masson.

Marks, Laura U. (2000), *The Skin of the Film. Intercultural Cinema, Embodiment, and the Senses*, Durham and London: Duke University Press.

Morrison, Bill (2006), "Retrospective," Cork, October 8–15, 2006: http://www.corkfilmfest.org/festival/bill-morrison.html (accessed November 14, 2014).

Paech, Joachim (1999), "Figurationen ikonischer n...Tropie. Vom Erscheinen des Verschwindens im Film," in Sigrid Schade, Georg Christoph Tholen, and Heiko Idensen (eds), *Konfigurationen. Zwischen Kunst und Medien*, 122–35, München: Wilhelm Fink Verlag.

Peirce, C. S. (1998), "What Is a Sign?," in The Peirce Edition Project (ed.), *The Essential Peirce. Selected Philosophical Writings. Volume 2 (1893–1913)*, 4–10, Bloomington and Indianapolis: Indiana University Press.

Rodowick, D. N. (1997), *Gilles Deleuze's Time Machine*, Durham and London: Duke University Press.

Schüller, Dietrich (1994), "Von der Bewahrung des Trägers zur Bewahrung des Inhalts", *Medium*, 4: 28–32.

Simmel, Georg (1959), "The Ruin," in Kurt H. Wolff (ed.), *Georg Simmel, 1858–1918. A Collection of Essays, with Translations and a Bibliography*, 259–66, Columbus: The Ohio State University Press.

Sitney, P. Adams (2002), "Structural Film," in *Visionary Film. The American Avant-Garde, 1943–2000*, 347–70, Oxford and New York: Oxford University Press.

Smither, Roger, ed. (2002), *This Film Is Dangerous. A Celebration of Nitrate Film*, Bruxelles: FIAF.

Stewart, Garrett (1999), *Between Film and Screen: Modernism's Photo Synthesis*, Chicago: University of Chicago Press.

Totaro, Donald (2004), "The Old Made New. The Cinematic Poetry of Bill Morrison," *Horschamp* (November 30, 2004): www.horchamps.qc.ca/new_offscreen/morrison_rebirthism.html (accessed November 14, 2014).

Usai, Paolo Cherchi (2001), *Death of Cinema. History, Cultural Memory, and the Digital Dark Age*, London: BFI.

Welch, Barnaby (2002), "Interview with Bill Morrison," *High Angle Magazine* on http://www.decasia.com/html/highangle.html (accessed November 14, 2014).

# 7

# Moving Images as Ontographic Images (According to Pier Paolo Pasolini)

Lorenz Engell

There can be only little doubt—and were it different, we would probably not contribute to a volume focusing on media and matter—that in the humanities, in cultural studies, and especially in media studies, for a couple of years now, a new interest in materiality and the material has come up. Some have already for years talked of a "material turn" in the humanities or of "material culture" as a special field of research and reflection (Bennett and Joyce 2010; Gell 1998; Latour 1991; Miller 2005). One of the questions that arise from this shift is to know where this recent interest comes from. Why is it that, for something like a decade now, we have, in the field of humanities, pursued research in matters and things so insistently? While in the 1980s, a programmatic concept of immateriality like Jean Francois Lyotard's "Les immatériaux" (1985) could be highly and prominently estimated, today we concentrate more and more on "les matériaux," on matters, even substances, and objects. When and how did this "material turn" happen?

But beyond this type of genealogic or archeological question, secondly and perhaps even more intriguing, we can notice that not only the mere fact of the coming up, this new field of research calls for explanation. The new interest in material and materiality does not so much aim at new theories or philosophies about matter and material or on materiality, but merely in those materials. It is not so much a reflection of materiality, but an observation of what is being produced in and by the material and the material world. As thinkers like Jane

Bennett and Karen Barad have shown, materially operating procedures occur within the material world itself. They equal reflections, at least in a functional manner, without claiming for some point or level of reflection beyond the immanence of the operating material itself (Barad 2003 and Bennett 2010). Christiane Voss hence suggests to replace the term of "reflection" to the benefit of the term "reflux," a bending and countering movement, which is produced in and by the flux of matter itself (Voss 2010: 178). Of course, we have to keep in mind that even such an observation presupposes a material body of technical, biological, or physical order, such as instruments or media. This is why this kind of thinking matter is so closely linked to thinking media (Mersch 2010: 196). This new "matteralism" (to discern it from traditional, conception materialisms of different kinds) does not only focus on the concept and conception of materiality, but also in the ontic matters themselves, in substances and in things of expression, of communication, of knowledge production, of sensemaking, of art. It does not so much ask for what matter or material is or may be, but for what matter or materials do, what they perform in producing and reproducing and organizing and knowing and finally thinking themselves. Hence, it does not inscribe itself so much in an ontological tradition, but it opens up a new field, which I shall call the ontographic field.

In the following chapter, I shall try to discuss both aspects at once, the archeological one and the ontographical one. I shall have, far below the most general and abstract level of debate I began with in the above paragraph, a closer look on one very restricted example both on the side of theoretical approach to matter and materiality, and on the side of the mediatic and material operationality and thinking of (technically and aesthetically organized and organizing) matter. Seen from the perspective of media studies in general, and especially from the perspective of media scholarship in Germany, as it has developed since the early 1980s, the turn toward the material grounds of communication is something like the founding paradigm of the discipline (Gumbrecht and Pfeiffer 1988). With the rise of media philosophy during the last ten or fifteen years in Germany, though, the issue of matter has to a certain extent been shifted. Media philosophy has to do with a sort of reconciliation of the philosophical tradition of conceptual work—or "*Begriffsarbeit*"—with the new material approach as it is predominant in media studies (Engell et al. 2013). Questions that had been denounced programmatically by media studies came back in media philosophy, such as problems of anthropology, or ethics, or ontology (Voss 2010; Engell 2014).

The example I want to have a closer look at comes from the field of film, and film theory. One of the reasons that make some even more specialized research into the philosophy precisely of film so promising in the realm of ontology/ontography is that an interest in ontic things and matters and in the ontology of this medium, the moving image, has always been present in film theory and aesthetics, and even foregrounded. It has, though, been suppressed during a specific period in the history of film thinking. So, within the framework of film philosophy, the question is less how the material could so unexpectedly appear in the late twentieth century, but why it had always been so central, and how it could ever disappear from the theoretical scene. This having the material of the moving image disappear was produced with the help of specific arguments, roughly between 1960 and 1980, and in parts far beyond until today, with the predominance of structuralist and poststructuralist, formalist and neoformalist, semiological and postmodernist approaches in film theory. These approaches turned the back on the problem of ontology in general, be it the question of the ontological status of the moving image itself, be it the question of the ontic world as being represented or reproduced or, as I will argue here with the help of Pier Paolo Pasolini, unfolding by means of the moving image.

In the following paragraphs, I shall try, with the help of this remarkable example, to reconstruct one of the most decisively material ontological approaches in film philosophy, which has at the same time been, maybe more than any other, subject to the suppression of the material ontology of cinema by rationalist and postrationalist theories from the beginning of the 1960s on, and in parts till today. Pursuing the very same course, I will at a time attempt to reconstruct this approach of ontology of the moving image in terms of more recent approaches as they are prominent in our debates today. By this I will try to open it up again and to show how today it could contribute to and participate in our debates on media and matter, which it was completely denied when it was first published. I am speaking of Pier Paolo Pasolini's more or less legendary, but in parts still undiscussed concept of film being, as Pasolini puts it, "the written language of reality," or the scripture of reality, or, as I would suggest to put it, an ontography. Film is, following Pasolini, a unique system of ontic, material devices, instruments, gestures, and operations by which material reality, or the ontic, writes itself (Pasolini 1972). This concept, which we will unfold more closely below, was first presented by Pasolini in 1966 at the film festival in Pesaro, Italy, and which was first much disputed in the discussion, then sharply rejected, and which has almost been ignored ever since.

In the tradition of film aesthetics, film theory, and film philosophy, there has, as I mentioned before, always been a very specific concern with the *ontological* status of the moving image. The famous question: "What is cinema?" has been asked, explicitly or not, over and over again (Bazin 1976). At the same time, film has always been researched as giving access to the world in a somewhat specific way, thus unveiling some special *ontic* quality of the world it opens up or, to be more precise, of what this image refers to, or produces in front of our eyes. In some cases, these two questions have even been put together, or piled up one upon the other. Since cinema, as a physical and technical device, as perceivable object to our senses, and as a system of material operations, is part of the physical and phenomenal reality, its own ontological status is at least in parts of the same nature as it is the case for the ontic world film opens up or generates.

In film theory, though, this quite rationalist idea of a strict separation between ontic referent and ontological signified competes with another and older one, developed in authors like Bela Balázs (2010), Jean Epstein (1921, 1926), Louis Delluc (1919, 1920), André Bazin (1976), Siegfried

Kracauer (1960), and, a little more recently, Stanley Cavell (1978), Gilles Deleuze (1985, 1989), and Bernard Stiegler (2001). According to them, there is precisely no sharp juxtaposition of what the material visible object of the outer world is and what would be the object as it is present on the film screen. The so-called pro-filmic, that which is in front of the camera as independently given, as just "being there," on the one hand, and the filmic, that what appears on the screen, on the other hand, are, to these approaches, linked by a specific relation. For Balázs, this relation is to be described as "visibility" (Balázs 2010), Epstein and Delluc call it "photogénie" (Delluc 1920), Siegfried Kracauer uses the term of "affinity" to grasp it (Engell 2008). In Cavell, the decisive term would be "becoming" (Cavell 1978): what becomes of things on film is his question, insinuating the idea that there is a continuous flow of material transformation that binds visually given things on film to the things just found by the camera.

Gilles Deleuze and Bernard Stiegler go even further. To the latter ones, writing decades after Pasolini, the world has always already been cinema, as Jay Hetrick stresses recently (Hetrick 2013). It is not the moving image that operates on a meta-level with respect to the world, delivering metaphors or metonymies of reality. It is rather the other way around, the universe is a metacinema, and it is precisely this idea that we can encounter much earlier in Pasolini, as I will show below. Anyway, these approaches, even if deferring considerable, share the assumption that there is at least an ontologically fuzzy or third space in between the pure image-object on the screen and the pure material thing in front of the camera. This fuzzy space is built up by specific interactions or operations that derive from the interplay of (1) the photo mechanic machinery of cinematography, as material and real as it is, (2) the material objects that surround this machinery and to which the camera is directed, and (3) the physical body of the spectator and the operations of perception and action that can be ascribed to this body (Voss 2006, 2013). At the very core, it is precisely the camera's pure and mechanical recording or registering of the light as it is refracted by the object in front of the camera, of the movement or gesture the object is subject to, and of the movements or gestures of the spectator. The automatic nature of the recording process, as underlined in Kracauer (1960) as well as in Cavell (1978) and in others, is responsible for the moving image's not creating or evoking mental images, or concepts, or signified entities in a strict semiological sense, but referents, objects in a full cinematographic materiality, which derive from the interplay of the signifying matter with the depicted matter in a deeply nonsymbolical and nonarbitrary way.

This *onto-ontological* property of the moving image can also be expressed in the well-established models of semiotics, or semiology, as they are grounding basics to quite a number of influential and relevant approaches in the field. If one translates the traditional ontological assumptions about the moving image into semiology, on the basis of the dichotomy of signifier and signified, they turn into a challenge, if not into major obstacle. The cinematographic signified as being represented by the signifier which the moving image is, has, according to traditional film aesthetics, a very specific relation to the referent, or the visible ontic object or movement in the outer material world it hints at or has recorded. In terms of an orthodox Sausurrean semiology, as it has, for example, been unfolded by Roland Barthes (1965) or by Umberto Eco (1968a, 1975), the signified is a purely mental image or concept, or, as for Umberto Eco, a "cultural entity" (Eco 1968a: 23). The referent, in sharp contrast, is a part of the material or real world beyond the sign, a physical object, for instance, and there is precisely no way in which the referent as a material object could ever participate in the being of the signified.

In Pasolini now, in a rather sharp short circuiting of semiological orthodoxy, the signifying matter of cinematography is even *identified* with the depicted matter or material world (Pasolini 1972: 81). If this were true, the ontology of the moving image would simply and plainly be the ontology as such. More than this, one could not even speak of an ontology of the moving image, since the gap between the ontic layer of reality and the conceptual (or semiological) layer on which the ontology of this reality could unfold and which would have to transcend the former one would be annihilated in the case of film. If the material world is already the signifying matter of the film, we will have to conceive of a more or less horizontal instead of vertical semiology, or ontology, a semiology, or ontology in which signs or concepts are no longer abstractions that range on some level above reality, but are to be found on the ontic level itself. They do no longer reflect the material world from outside, or from above, but form a part of it. Moving images do, according to this approach, not write upon or about reality, but, as being real matter in themselves, they write reality itself by reality. To be more precise, moving images are the matter-reality of matter-reality describing or writing itself. They are—ontographic.

Notwithstanding the juxtaposition of his ontology and the principles of semiology, Pasolini displays his concept of film as written language within the conceptual grids of then contemporary film theoretical debates, which were mostly influenced by contemporary semiology and structuralism (Pasolini 1972: 72). Moreover, he did so under the influence of the then extremely important and influential essay of Christian Metz: "Le cinema—langue ou langage," thus placing himself in the realm of Saussurean, structuralist semiology (Pasolini 1972: 75–8; Metz 1964). Using these terms and concepts that were very fresh and avant-garde in film theory at that time, though, gave way to the complete turning down his ontology, because both operate on completely different epistemological and philosophical grounds.

In his text, Pasolini refers to Christian Metz's famous model of a possible grammaticality of film. Film, says Metz, knows rules and orders and structures on the level of individual utterances, comparable to the spoken language, or "*langage*," but it does not have an abstract conceptual grammar on the level of the language system, or "*langue*" (Metz 1964: 86). In verbal language, Metz says, a sharply restricted set of elements—like the set of phonologically relevant sounds of a specific language—and rules of combination could produce an endless number of different forms (Metz 1964: 58). In film, on the contrary, we have, according to Metz, to do with an endless, and arbitrarily composed, open set of elements, which are the possible images that a camera could capture and which are innumerable (Metz 1964: 67). And also the rules of combining images are, in principle, not restricted in number and diversity. Nonetheless, and surprisingly, departing from this more or less formless situation, feature films on a more complex level form a strictly systematized and sharply restricted structure of possible individual utterances, formally described by Metz in his "Grand Syntagmatic" of film (Metz 1966).

Pasolini now contradicts both parts of Metz's basic assumptions (Pasolini 1972: 75–8). The elements of a possible film language are neither arbitrary nor undefined. On the contrary, the basic isolated elements of film languages have always already been well defined by human and social and technical practices and procedures. The elements of film language are the objects and the actions or gestures or operations of reality themselves (Pasolini 1972: 75). The elementary bodily gestures and actions of humans and the material objects they relate to and from which they derive do constantly work, and re-work, and hence produce, reproduce, and change ontic reality. They are as real as any reality could be. Since, though, at the same time, they are the basic and most primary material of human expression, or expression of any kind, they are to be seen as the root of any system of signification or language or any sign system. The objects and operations of reality are always already grammatical, says Pasolini (1972: 77). Acting, or having agency, means, as Pasolini puts it, always already writing a poem, the most relevant and beautiful of all would be, of course—for the Marxist Pasolini is—the revolution. Hence his formula about film as scriptural form of human action, or film as "lingua scritta della realtà."

The real, ontic objects and operations now are, as Pasolini puts it, by no means being represented or imitated by the moving image. They are being reproduced in another matter, the matter of the celluloid-based projected image. Even to speak of "transformation" here would not be completely adequate, since Pasolini's concept is more one of dislocation or transportation of the object from one place—we might call it the scene of action—to another place, we might call it the screen

(Pasolini 1972: 82). The movement of the object from scene to screen would, for Pasolini, not be a vertical one, one that generates a "higher" or more abstract, pictorial or ontological level, but a purely lateral one. The elementary ontic objects and gestures become film, and again we have here an interesting precession to the use of that very formula by Stanley Cavell, who presumably never read Pasolini's essay, some ten years later (Cavell 1978). In becoming film, the objects and gestures nonetheless keep at least something of what they are; they do not separate strictly from their ontic being. As Pasolini metaphorizes, the camera operates like a fishing device, it fishes the elements of reality, such as objects, and gestures, and operations, holds them back, wraps them, and incorporates them into film (Pasolini 1972: 82).

In a more precise quasi-technical description, Pasolini identifies a whole set of operations during the production of a film that result in this incorporation or embodiment of reality into film, like setting light on the scene, like building the site or the set of action, like selecting the position and angle of the caption and the length of the shot, like editing, etc. (Pasolini 1972: 82–8). This part of Pasolini's essay today seems rather conventional and rudimental, with the remarkable exception of Pasolini's insisting on timing as a crucial operation (Pasolini 1972: 88). It is through timing that the material operations in front of the camera, the operations of the moving image itself, and the movements of and in spectators in the movie house are interconnected in one coherent line of transformation or literally transportation. Wrapped in time, once turned into cinematographic signs, even as signifiers, the objects and gestures do not stop being what they are, or at least not completely. This is why, says Pasolini, cinema does not need a specified semantics, since the objects of the material world, or referents, are always already wrapped into the significant matter of the celluloid image (Pasolini 1972: 85–6). The reality of elementary objects, and operations, is the reality of cinematographic signifiers and hence is a momentum of the language of film. If one looks at this concept of Pasolini's more closely, it seems to result in negating the signified. Therefore, there is, for Pasolini, neither a semantic nor an ontological difference between the moving image and its outer or depicted reality. It is exactly this assumption that fiercely scandalizes Umberto Eco since it results in nothing else than the destruction of the concept of the sign as a purely conventional and cultural entity of strictly intellectual and abstract character (Eco 1968a, b).

Today, especially in the framework of epistemologies and ontologies of acting networks, inscriptions, cycles of reference, Pasolini's ideas may be worth a reinspection (Latour 1991). They could again figure in theories of distributed agency in film and of film as specific form of an "acting field" (Engell 2010). According to Pasolini, the objects, the gestures, and the operations that make up the language of film by becoming film, are by no means the isolated elementary entities as they are so dear to structuralism, but they always form, as we might say, assemblies (Engell 2010). They may be distinct, but never isolated, they form complex fields from which they cannot be extracted, in which they not only interrelate, but also produce, and reproduce, and change each other. These fields are at the same time made up by the objects and operations, and conditions of their possibility of being.

And what is true for the cinematographic caption or production is, according to Pasolini, relevant also for its reception or projection (Pasolini 1972: 79–80). Film is not to be read or to be decoded by its spectators; no meaning is to be isolated from it. Instead, film produces a reality while being projected, namely, the reality of affects, of percepts, of emotions, of impulses and impacts and ideas whose material support are the spectators. Their movements and operations and gestures in the most different fields of action can be seen as the utmost transportation to which the process of film as writing reality may reach. The reality of the moving image may, according to Pasolini, by its affecting force and impact, be reinscribed into reality when it comes to motivate action and reflection beyond the movie theater (Pasolini 1972: 80). Here, Pasolini's concept could be lined up with the most recent tendencies and interests in film philosophy that have turned more and more from questions of reflexivity and thought to those of affectivity, of attachment, of binding and coupling forces, and of emotionality (Voss 2013; Gregg and Seigworth 2010). More than anything else, these escape from the rigidity of semiological and postsemiological theory and criticism. To Pasolini, the language of film is, such as it is true for poetry, per se a form of action, not separate from reality, but emerging from it, transforming it, and performing the very same claim for relevance.

As appealing and interesting these assumptions and positions of Pasolini's are in film philosophical contexts of our days, as for example in the debate on media and matter we have here on this conference, as counterintuitive and awkward they must have appeared to the eyes of the developing and unfolding debate in film theory of the mid-1960s, which was just beginning its journey into deep semiological orthodoxy and its unfolding. Due to its deeply anachronistic character, Pasolini's paper underwent, at least in the German context, a strangely displaced and somewhat incomplete reception. On the one hand, it served as (counter-)point of departure to one of the most influential enterprises in the film theory of the late 1960s and early 1970s, namely, Umberto Eco's undoubtedly ingenious attempt to ground and place film as a sign system within the Saussurrean semiology, ascribing to it some exclusively special feature, which Eco called the "triple articulation" in the sense of Martinet's hypothesis of the "double articulation" of verbal language (Martinet 1949). For the reception of film semiotics in Germany, Eco's text was translated and published first in 1968 in a journal named *SPITZ*, and again in 1972 as a chapter in the German version of his "La struttura assente," also from 1968—in German "Einführung in die Semiotik." These two versions of Eco's text, the article and the chapter version, present the same central concept of "triple articulation," but they differ in their structure and in some text passages. In both versions, though, Eco refers openly and at length to Pasolini's concept of cinema as the written language of reality as developed in the Pesaro paper quoted above.

While the essay version, though, treats Pasolini's text with at least some polite analytical distance and critical respect, the chapter version rejects it in an overtly polemic and very sharp manner as contrary to any semiological reason. Pasolini's idea, says Eco here, is a unique semiological naiveté and contradicts even the most elementary and most undisputed findings in the theory of signs (Eco 1968a: 251). And it is not difficult to accept that Eco is completely right in this respect. Seen from today, the result of this double attempt to shape Pasolini's most radical materialist vision with the help of the semiology of the day is a more or less cumbersome, but also, in the best way, anachronistic theoretical design. Basically, it takes an already postsemiological position, insofar as it argues, in some central assumptions, beyond the logics of representation. Pasolini conceives of the cinematographic or filmic sign explicitly not in terms of a means of representation, and also not as evocation of an impression of reality in the perception or in the consciousness of a spectator, as, for instance, a phenomenological approach, as opposed to semiology, or the apparatus theory would do. Hence, the assumption of a founding arbitrariness of the cinematographic sign, so central and essential for a semiological analysis, has no function in Pasolini. Pasolini does not go into questions of indexicality and iconicity of the moving image, either, nor discusses the problem of the production of meaning or sense by and in the moving image, nor examines the functioning of filmic narration, as did most of the theoretical and analytical approaches in the realm of semiological film studies and theories.

Even more, Pasolini explicitly rejects the basic idea of general semiology, especially strengthened by Roland Barthes (1965), according to which verbal language were the most general and obligatory fundamental model in the light of which all other systems of signs and cultural production were positioned and hence had to be researched as languages. On the contrary, Pasolini concentrates almost exclusively on what one could describe today as the operative potential or even the agency of the moving image, its character as action, as setting up reality, as what I would suggest to call "ontographic," and hence as ontic reality itself, or constitutive and operative part of it. To my knowledge, this notion has been brought up by Graham Harman, and by Ian Bogost, in the context of what to them is a noncorrelationist, posthuman ontology, which is, basically, an ontology without a (privileged) human consciousness as its support or central reference (Harman 2010: 124–35; Bogost 2012: 35–59). To them, nonhierarchical, nonsemantic, nonsymbolic, and, interestingly enough, nontimely inscriptions of reality are "ontographic," such as lists, or diagrams, or photographs (Bogost 2012: 38). Film, at least to Bogost, is explicitly excluded from this concept of ontography for it being time based. Hence, any kind of changing material reality is excluded from this kind of "ontography," but it is exactly his interest in time, and change, that grounds Pasolini's and the moving images ontography (Bogost 2012: 47–50).

The ontic reality of film is, according to Pasolini, not following nor preceding any outer reality of completely different nature, and one does not represent the other one, either. Instead, both realities contribute to one another, inscribe themselves into each other. They are always already referring to each other, and, since they are inseparably interrelated, they form just one single reality that one could conceive of as "intra-active," as Karen Barad does it for physical reality (Barad 2003: 83–93). Cinema, writes Pasolini, is the written language of the reality of action, but this reality of action, on its turn, is always already cinema. It is very intriguing that this idea of the world, even before and without cinema, having always already been cinema, comes explicitly, and precisely in these very words, back in film philosophy with Gilles Deleuze's two volumes on the topic (Deleuze 1985, 1989). But Pasolini's approach is relevant far beyond the realm of cinema. Read in the light of Karen Barad's ideas about "Agential Realism," we can see the close parallel between Barad's and Pasolini's critique of semiology, and of discourse-centered thinking (Barad 2003: 7–10). Even more striking is the idea of matter acting upon matter via a chain of cuts and bends, and by this creating what we may call "facts" (Barad 2003: 78–80). Barad concentrates exclusively on the reality and onticity of scientific research, of the cutting off of laboratory situations, and on scientific "facts"; but, with Pasolini, we can as well apply these observations to the cutting off of the pro-filmic on the set of film production, the creation of moving images as aesthetic "facts" and of the gestures induced by those images as social "facts."

Notwithstanding this, on the other hand, Pasolini at the same time, and again explicitly, is looking for a possibility to ascribe a set of rules, or a grammar, to this writing its own reality of reality by means of the film; and he finds it precisely in the terminology of then recent semiological debates. And these terms, even if he has to distort and to misuse them, help him to describe and to analyze the processes of mutual inscription of film and reality of action more closely and more precisely, as he hopes. By which observation I want to point out that, even if semiology was the wrong friend, it was necessary or at least helpful to Pasolini in order to shape and to get across his concept of ontography of cinema.

Eco's own and complete theoretical intuition and intention and effort at that time were devoted to enforce the strict and infinitely deep separation between nature and culture, between the "real" material world and the world of signs, between the substantial and the purely differential, between two, and in the case of the moving image of three distinct levels of articulation, between the signifier and the signified, and the signified and the referent, and, of course, between the ontic and the ontological (Eco 1975). Even in the case of iconic signs, qualified by Charles Sanders Peirce as signs of resemblance, sharing, and overlapping between the sign and its object (Peirce 1903), and hence in the case of photographic and cinematographic images, Eco insists firmly on this point and establishes the iconic as a nonnatural, but completely cultural regime (Eco 1975).

Pasolini, on the other hand, as we have already mentioned, is interested precisely in the mutual generating and inscription of the respective realms. How is it, asks Pasolini, that from material reality a moving image can emerge, how could reality grow into film, and how could film reach back into reality or dissolve again into it. Which are the procedures of transformation between the moving image and its seemingly outer sphere of real human action? How conceiving of film as literally taking place within one single reality, which on its turn is to be thought of as being coherent, but dynamic, as being subject to constant change and last, not least as comprehending both the moving image and the pro-filmic given objects and operations at the same time? Pasolini's questions, other than Eco's, are, even if they are developed and treated as semiological problems by Pasolini, not of semiological descendance. These are questions of a possible operative ontology of film, expressed or disguised in semiological terms, or, to be more precise: these are again questions of a possible description of the moving image as being always already an ontographic apparatus and process.

In his later writings, Pasolini never came back explicitly to this concept of "lingua scritta della realtà." While other theoretical findings of Pasolini's, such as his concept of the free indirect subjectivation in film (Pasolini 1965), knew a broader and positive reception, his ontography of the moving image stayed more or less isolated, even if it was, as we have already seen, paralleled or reinvented later by authors like Cavell, Deleuze, and Stiegler. Pasolini was killed in 1976. Some twenty years later, his main opponent, Umberto Eco, though, began to notably shift his position, even without ever mentioning Pasolini again. In his critique of resemblance, Eco interestingly withdraws his earlier rejection of any possible bridging between the referent and the signified (Eco 1998). And in his second novel, *Il pendolo di Foucault*, Eco explores at length a possible world, which could be our world, which is always being written, producing itself while writing itself, this time not cinematographically, but literally literally, on a small sheet of paper (Eco 1988). Taken as a symptom, this shift of Eco's epistemology and ontology may be read as one of the signals of the turn toward materiality in the humanities. In the light of Pasolini's essay, though, this turn could as well be looked upon as an early indicator for the advancement of a specific strain of material thought, such as Barad's "agential realism," which is contrasting the language-centered thought we have to do with not only in semiology but in vast fields of the philosophical tradition as well. We allow ourselves, with regard to Pasolini, to qualify it not only as an ontographic, but also as specifically cinematographic thought.

# References

Balázs, B. (2010), *Early Film Theory. "Visible Man" and "The Spirit of Film"*, ed. Erica Carter and trans. R. Livingstone, Oxford, UK, New York, NY: Berghan Books.

Barad, K. (2003), "Agential Realism: How material-discoursive Practices Matter," *Signs*, 28 (3): 803–31; German: *Agentieller Realismus. Über die Bedeutung materiell-diskursiver Praktiken*. Aus dem Englischen von Jürgen Schröder. Berlin: Suhrkamp, 2012.

Barthes, R. (1965), "Éléments de sémiologiem," *Communications* 4 (10): 128–71; Engl: *Elements of Semiology*, trans. A. Lavers and C. Smith, New York, NY: Farrar, Straus & Giroux, 1997.

Bazin, A. (1976), *Qu'est-ce que le cinéma?* Paris: Cerf; Engl: *What is Cinema?* trans. H. Gray, vol. 1, 2, Berkeley: University of California Press, 2005.

Bennett, J. (2010), *Vibrant Matter. A Political Ecology of Things*, Durham, NC: Duke University Press.

Bennett, T. and Joyce, P. (2010), *Material Powers: Cultural Studies, History and the Material Turn*, London: Routledge.

Bogost, I. (2012), *Alien Phenomenology, or What It's Like to Be a Thing*, Minneapolis, London: University of Minnesota Press.

Cavell, S. (1978), "What Becomes Of Things On Film," *Philosophy and Literature*, 2 (2): 249–57.

Deleuze, G. (1985), *Le cinéma, t. 1: L'image-mouvement*, Paris: Minuit; Engl: *Cinema 1: The Movement Image*, trans. H. Tomlinson and Barbara Habberjam, Minneapolis: University of Minnesota Press.

Deleuze, G. (1989), *Le cinéma, t 2: L'image-temps*, Paris: Minuit; Engl: *Cinema 2: The Time Image*, transl. H. Tomlinson and R. Galeta, Minneapolis: University of Minnesota Press.

Delluc, L. (1919), *Cinéma et compagnie*, Paris: Grasset.

Delluc, L. (1920), *Photogénie*, Paris: de Brunoff.

Eco, U. (1968a), "Il codice cinematografico," in *La struttura assente*, Milano: Bompiani, 232–48; German: *Einführung in die Semiotik*, 250–66, München: Fink, 1972.

Eco, U. (1968b), "Die Gliederungen des filmischen Code," *SPITZ*, 27: 32–50.

Eco, U. (1975), *Trattato di semiotica generale*, Milano: Bompiani; Engl: *A Theory of Semiotics*, Bloomington: University of Indiana Press.

Eco, U. (1988), *Il pendolo di Foucault*, Milano: Bompiani.

Eco, U. (1998), "Réflexions à propos du débat de l'Iconisme" (1968–98), *Visio*, 3 (1): 9–32.

Epstein, J. (1921), *Bonjour cinéma*, Paris: La Sirène.

Epstein, J. (1926), *Le cinématographe vu de l'Etna*, Paris: Les Écrivains réunis; Engl: "The cinema seen from Etna," in *Jean Epstein: Critical Essays and New Translations*, eds S. Keller and J. N. Paul, trans. S. Liebman, Amsterdam: AUP.

Engell, L. (2008), "Affinität, Eintrübung, Plastizität. Drei Figuren der Medialität aus der Sicht des Kinematographen," in Stefan Münker, Alexander Roesler (eds), *Was ist ein Medium?* 185–210, Frankfurt (M): Suhrkamp.

Engell, L. (2010), *Playtime. Münchener Film-Vorlesungen*, Konstanz: UVK.

Engell, L. (2014), "Ontographie und Anthropogenese," in L. Engell, C. Voss (eds), *Mediale Anthropologie*, München: Fink.

Engell, L., F. Hartmann, C. Voss, eds (2013), *Körper des Denkens. Neue Positionen der Medienphilosophie*, München: Fink.

Gell, A. (1998), *Art and Agency: An Anthropological Theory*, Oxford: Clarendon Press.

Gregg, M., G. J. Seigworth, eds (2010), *The Affect Theory Reader*, Durham, NC: Duke University Press.

Gumbrecht, H. U., K. L. Pfeiffer, eds (1988), *Materialität der Kommunikation*, Frankfurt/M: Suhrkamp; Engl: *Materialities of Communication*, Stanford: Stanford University Press, 1988.

Harman, G. (2010), *The Quadruple Object*, Winchester, UK; Washington, DC: Zero Books.

Hetrick, Jay (2013), "Cine-esthetics: A critique of Judgement after Deleuze and Michaux," PhD diss., University of Amsterdam, Faculty of Humanities.

*Les Immatériaux* (1985), Paris: Centre Georges-Pompidou.

Kracauer, S. (1960), *Theorie des Films. Die Errettung der physischen Wirklichkeit*, Frankfurt (M): Suhrkamp; Engl: *Theory of Film. The redemption of physical reality*, Oxford, UK: Oxford University Press.

Latour, B. (1991), *Nous n'avons jamais été modernes. Essai d'anthropologie symétrique*, Paris: La Découverte; Engl: *We have Never been modern*, trans. C. Porter, Cambridge, MA: Harvard University Press, 1993.

Martinet (1949), "La double articulation linguistique," *Travaux du Cercle Linguistique de Copenhague* 5: 30–7.

Mersch, D. (2010), "Meta/Dia. Zwei unterschiedliche Zugänge zum Medialen," *Zeitschrift für Medien- und Kulturforschung* 1 (2): 185–208.

Metz, C. (1964), "Le cinéma: langue ou langage?" *Communications*, 3 (4): 52–90.

Metz, C. (1966), "La grande syntagmatique du film narrative," *Communications*, 5 (8): 20–4.

Miller, D., ed. (2005), *Materiality*, Durham, NC: Duke University Press.

Pasolini, P. P. (1965), "Intervento sul discorso libero indiretto," *Paragone*, 21 (6): 121–44.

Pasolini, P. P. (1972), "La lingua scritta della realtà," in Pasolini, P. P: *Empirismo Er etico*, Milano: Garzanti; Engl: "The written language of reality," in Pasolini, P. P. (ed.), *Heretical Empiricism*, eds B. Lawton, Washington, DC: New Academia Publishing 2005; German: "Die Schriftsprache der Wirklichkeit," trans. M. Cuntz, *Zeitschrift für Medien- und Kulturforschung*, 3 (2): 76–90.

Peirce, C. S. (1903), *A Syllabus of Certain Topics of Logic*, Boston: Alfred Mudge and Son; German: *Phänomen und Logik der Zeichen*, eds. Helmut Pape, Frankfurt (M): Suhrkamp, 1983.

Stiegler, B. (2001), *La technique et le temps. Tome 3: Le temps du cinéma et la question du mal-être*, Paris: Galilée.

Voss, C. (2006), "Filmerfahrung und Illusionsbildung. Der Zuschauer als Leihkörper des Kinos," in Voss, C., Koch, G. (eds), … *kraft der Illusion*, 71–85, München: Fink.

Voss, C. (2010), "Auf dem Weg zu einer Medienphilosophie anthropomedialer Relationen," *Zeitschrift für Medien- und Kulturforschung*, 1 (2): 169–84.

Voss, C. (2013), *Der Leihkörper. Erkenntnis und Ästhetik der Illusion*, München: Fink.

# 8

# Brain Matter and New Phrenologies: Challenging Brains with Melancholy and Vice Versa

Benjamin Betka

In Chuck Palahniuk's short story *Zombie*, young brains flicker and expire. Promising students of American high schools combust under the pressure to succeed and make life-altering decisions: they obtain their school infirmaries' defibrillators across the nation to give themselves a shock treatment to the temples (Palahniuk 2013). They reduce their IQs instantly and enable themselves to live a more simple life: no longer are they capable of minding an auspicious but risky future. It is a halfway suicide. They eagerly change themselves into *less vivacious* beings and become *undead*. It is a liberation through electric stultification. Brain matter gets rectified and simplified. Young learners say no to initiative and mature thought. The short story can be read as a bleak satire on the perpetual rise of brain enhancement with molecular means. While drugs such as Prozac and Ritalin are tools to fit into contemporary consumer industry, the voluntary shock therapy here is a desperate measure to escape it. In both cases, the brain is the central site, the battleground even, and *neuronal* conditions matter here.

Thinking, one of the brain's activities, can be homicidal and fatal. Especially in the academic field, strong minds are welcome and frequently associated with happiness, superiority, or even progress and salvation. Roland Barthes noted the auratic qualities of Einstein's brain in his *Mythologies* (2003: 24; see also Pepperell 2003: 14). But mental work is not *necessarily* productive or positive. In a society that no longer depends on manual craft but intellectual labor, the pressure between the ears is rising—which Palahniuk's raucous story illuminates harshly. The brain is more than the producer of solutions, leading toward prosperity and there is something like *over*thinking, fruitless rumination, and unframeable but persistent doubt and skepticism. It can keep its host (or owner) from sleep, joy, and participation, thereby making him or her a victim of thought. It is a state of melancholy, depression, and fear that led to fatal decisions as long as this species existed (Jamison 2000: 11). A society has to treat these cases as deplorable exceptions from the norm but for the afflicted themselves, their state of mind is not an exception but the inescapable and ubiquitous base for their conscious existence.

The following can be understood as going *against the brain*, or at least in sowing doubt on the usefulness of a too simple, unequivocal appreciation of this (undoubtedly fascinating but eventually feeble) nervous tissue alone. It goes against the brain that is only good and interesting and mind-opening. The notion of *plasticity* promises a deeper look at this firstly scientific and much hyped object that is the brain, but it does not automatically do its complexity justice. Any anatomistic, segregating tendency of neuroscientific endeavors must be countered: any useful theory of the brain has to include its capacity to inflict self-harm, to maim, and to kill. Like a starving organism begins to digest its own organs, the brain is a potentially hazardous matter that can turn against itself.

## One: Brain growth and sparks in context

The brain became an explanatory instance and/or point of reference for about everything during the last few decades as it was included into diverse clinical, academic, and popular discourses (Sacks 2010; Sturma 2006). The new imagery of synaptic tissue enabled new negotiations of traditional problems such as the mind|body axis and the framing of evasive concepts such as attention, thought, and consciousness. But first and foremost, the brain is nothing but one *concept* among others in the scholarly field. In the translator's foreword to *A Thousand Plateaus— Capitalism and Schizophrenia* Brian Massumi writes: "A concept is a brick. It can be used to build the courthouse of reason. Or it can be thrown through the window" (2004: xiii). The brain offers itself as such a big brick: Can this not be the authority that finally explains to mankind *what* and *how* it is? Massumi mutes any enthusiasm quickly as he continues: "Because the concept in its unrestrained usage is a set of circumstances, at a volatile juncture. It is a vector: the point of application of a force moving through a space at a given velocity in a given direction. The concept has no subject or object other than itself" (2004: xiii). The brain only sits at the heart of things (no pun intended) because a scholarly community puts it there. Alternative conceptualizations are possible and needed when confronted with melancholy's scope.

The concept of a central brain has a certain lure. However, skepticism is in order as this new imagery also brings a renewal of humanist self-centeredness that narrows the neuronal paradigm. Frequently, explorations of the brain are philosophically unreflected. It is not the brain as a central entity that helps to encounter this picture but rather life as a whole: as one does not question what the brain *is* but what it *does* one is led to its components. Donald Hebbs's theory says that neurons that fire together, wire together. In 1949 Hebb enabled the basics for synaptic *plasticity*—the ability of synaptic tissue to become more or less complex over time according to the degree of its activity or participation in larger circuitries (Hebb 1949). The spark, the sudden firing, is thus responsible for material expansion—which yields different firing patterns. The spark bears resemblance to the theoretical notion of a concept itself: it is also no subject or object other than itself. It is an event but not an item: it alters objects and structures just as it depends on them to be relevant. The lure persists in the notion of identity and unity as the synapses have the potential to form something *bigger*.

However, a too simple notion of plasticity bears an unreflected allegiance to positivism and progressivism. A more wary and thus more encompassing and recalcitrant perspective is useful to avoid *new phrenologies*, the myopic idealization of a (howsoever complex) chunk of fleshly matter. This means to leave more traditional terrains, the ones that argue for a stable and more or less responsible and response-able human being in*corp*orating this brain and its firing, wiring cell heaps— *e pluribus unum*.

Hebb's slogan has a practical use as it introduced a temporal dimension into pedagogics. It supports the idea that one can actually *make* brains, or at least have an influence on what and how it happens within the tissue. This has emancipative potential—but it can also be seen as providing the basis for human engineering which spreads specific general principles. The humanist ideal that keeps aiming for something "better" shapes, but can also obstruct, the view here. In the case of Palahniuk's zombies, institutionalized brain growth resulted in their voluntary amputation from autonomous existence. Discussing the brain and its many conceptualizations as bricks or projectiles frequently touches upon grander current horizons: today's treatment of the brain is significant for a bigger paradigm shift in Western culture as Robert Pepperell asserts in *The Posthuman Condition: Consciousness Beyond the Brain*: "We are nearing an awareness of the energy of existence— there is the tangible crackle of a storm in the air" (2003: iv). Speaking of the brain and the assumption that it somehow *houses* consciousness as it *houses* these synaptic tissues, Pepperell continues: "It is a commonly held belief that the brain determines or causes mental phenomena, in particular the phenomena of consciousness, with the consequence that in much philosophical discussion of consciousness the body and the world beyond are largely neglected" and that "if we are to make any progress in understanding the role of consciousness in our existence" we must avoid this neglect (2003: 13).

## Two: Naked philosophy and neurophilosophies

The neuronal dimension grows in scope once the melancholic disposition is considered and thus is not merely categorized as of a

mentally somewhat obstructed (but curable) minority. A stellar imagery
for overthinking and restless rumination in Western civilizations can be
found in Dürer's *Melencolia I* from 1514, a long time before the advent
of neuroscience (Klibansky et al. 1992: 406–48). The image is the result
of a synthesis of dominant forms of expression of the melancholy, which
the artist combines (Klibansky et al. 1992: 448). On this copperplate,
the figure[1] can be understood as being on a severe inquiry for the *sum*
of things, for the biggest frame of relevance. This transcends cheering
curiosity but denotes a rather exhausting enterprise of solitary mental
work. She is neither agitated nor sleepy. The slumped figure in the image
ignores all the interesting and probably useful tools and items scattered
around. They remain unused resources. She might seem cherubic, but
the image deals neither with grace, benevolence, nor with a promise
of salvation. Her gaze is blank and aims at some distinct horizon and
the face is darkened. The head is heavy and has to be supported by
the left arm, while the hand is clenched in a fist. The image combines
inactivity and tension, isolation and containment but also suggests
ongoing introspection—and retrospection. Where is *Melencolia*? She
sits *in-between*, neither fully detached nor anchored.

   A fundamentally philosophical quest is depicted here, five centuries
ago—a quest that can have tremendous weight and serious results.
Indeed, one could understand melancholy as *primal* or *naked* philosophy,
as unsheltered and risky in its seclusion: it is a deep pondering of words,
meanings, relevances, wholes, parts, and, above all, *the daily routines
of understanding* that a majority might share. It does not have positive
connotations and surely is not enjoyed: How did it start? It was sparked
not by a defibrillator but by a shock. Deleuze doubts "that thinking is
the natural exercise of a faculty, and that this faculty is possessed of a
good nature and a good will. 'Everybody' knows very well that in fact
men think rarely, and more often under the impulse of a shock than in
the excitement of a taste for thinking" (1994: 132).

   It means to mind the past and the future alone, thereby detaching
from the concerns of the present tense. This mental work associated with
uncompromising philosophy and the rigorousness of melancholy is not

(yet?) transformed into an institutionalized and stabilized discourse: the latter is reserved to philosophy departments and libraries around the world (which are sheltered, and hold on to a specific discourse and dress code). The process of melancholic delve does not depend on books; it is not communicative or educative. As a thinking process that cannot be shared, it is not interdisciplinary but only *pre*-disciplinary in the academic and any other context: Klibansky et al. argue that with Dürer's image it becomes possible to understand melancholy as *beyond* the simple dualism of health and sickness (1992: 490).

Melancholy is hard to conceptualize and cannot be isolated, unless in a clinical|medical discourse where norms are established and defended. It aggravates the problem of the observer in system theory: the melancholic point of view is totalitarian and a neutral assessment of what is seen is not possible. Once triggered by a shock of any kind, melancholy is inescapable and absolute—just like the plasticity of the brain, just like consciousness. It is both exclusive and inclusive and cannot be altered instantly on a mental level alone, only radically, as Palahniuk's students show.

Questioning values and specters of the future as a whole is reserved to the liberal arts and the humanities in which an academic uttering can be easily understood as the result of a larger zeitgeist. As scholarship begins to acknowledge the brain more and more, a fundamental problem of academia in the West comes into view. It is older than MRIs, Prozac, and zombies in the United States, as Schwartz and Begley summarize:

> [The] division of mind and matter was ... something of a scientific debacle. Separating the material and the mental into ontologically distinct realms raised the white flag early in the mind-body debate: science abandoned the challenge of explaining how the components of the physical world found expression in the mental world. And thus was Cartesian dualism born. Today, three and a half centuries later, his belief endures. (2003: 33)[2]

As a concept and as an idea it is ancient, sprawling, and impossible to frame in a singular scholarly study for its interdisciplinary width

(Klibansky et al.; Burton; Solomon). Today, as posthumanism unfolds, one is to consider melancholy and other states of consciousness *not* as an exception from the norm, thereby struggling with the observer's dilemma: the condition bears a name but the condition*ed* are not versed to put it into context.[3]

A fundamental disposition lies in the challenging word *neuro* that conquers others when attached to them: neuroscience, neurophilosophy, neurohygiene, neuroculture. Does not every morpheme that gets attached to another cause a reduction, a specialization, a narrowing of meaning? The longer the compound word, the more specific it is. When the brain and the entire nervous system become the center of attention, does one not simply bypass fundamental and still unresolved philosophical questions by doing so? This is more than wordplay. With *neuro*, a new materialism seems to enter the academic circuitries: a favored and privileged kind of matter subordinates other theories. On the terminological level this means: Why bother with abstract *-ologies* (epistem-, ont-, phenomen-, etc.)? Why not find the newly detectable patterns in the brain first and jot down the implications for individual performances and biographies? Dieter Sturma writes in the introduction for a collection of papers discussing synaptic and other matters that there is even a disappointment and chagrin regarding the Arts in this beginning neuro-age (2006: 7). A new conception of the human being and its conditions seems to be in the works—the body-mind problem, which ancient and modern philosophies kept churning without ever reaching a majoritarian solution, is reformulated as the question for the relationship between psychological states and physical phenomena. According to Sturma, neuroscientific and neurophilosophical positions claim that this relationship will soon be defined for good and that the humanities are about to take over mere translation work or offer illustrative incidents for the sciences (2006: 9).

Along these lines, *neuro* seems to oppose melancholy (and/as philosophy): the former is integrative, meticulously humanizing, *zooming in*; the latter disposition is segregative, widening horizons, *zooming out.* A frustrating polemic arises: one could think of the term *neuro* as a drill head the schools of science aim toward the vast bulk of the humanities in order to take over.

## Three: Brains as centers, products, and participants

Can the philosophical and the (neuro-)scientific points of view be joined in a more synergetic manner, and not in a unilateral invasion? Is there a common ground to examine and appreciate the brain and its supposed products which we call *experience* or *thought*? Stephen Rose writes in *The Conscious Brain* that it "is biology's greatest challenge. Perhaps ... it is the greatest challenge for science as a whole ... each [one] is two fistfuls of pink-grey tissue, wrinkled like a walnut and something of the consistency of porridge" (1978: 21). Within this mediating matter, "[received] data is transformed into a series of electrical signals passing along particular nerves to the central brain regions where the signals interact with one another" (1978: 23). Rose does not write that the brain is philosophy's greatest challenge but implies that philosophical questions will eventually appear as well. He maintains eventually: "Nonetheless in some sense we are all solipsists; the external world is and reinterpreted through our mind's eye. ... And when dealing with other objects as significant for us as those we interpret as 'fellow humans' we view their actions in particular from a point that owes much to a solipsist perspective" (1978: 25). He highlights the brain as giving the ability to speak (1978: 175). Speech creates speakers and while Rose does not ponder the weight of personal pronouns explicitly, he quickly enmeshes the image of a single, unique person into his remarks. An affirmation of individualism is also made when he discusses the brain's plasticity in contrast to its specificity. "Specificity may lay down the equivalence of identical twins, but plasticity ... makes each the sum of his or her own unique experiences" and "if the brain were not plastic in this sense individual humans would be almost totally and comprehensively programmed, like ants or bees" or zombies (1978: 212). Plasticity is the central aspect of being human and leads to a personality. The Enlightenment shines here: the divine watchmaker has put all the pieces into place so they can perpetually coalesce.

Another way to discuss plasticity would have been possible (but not likely) here: plasticity *could* have been called a persistently disturbing factor that makes human individuals so unnervingly diverse. While it then could be admitted that plasticity gives human beings an evolutionary advantage that enables the species' survival, the constantly irritating need for communication and identification would be lamented. Rose endorses individualism while not actively calling for it. The brain thus becomes an agent on a frequently unquestioned person-centered agenda.

This perspective can be complicated by considering the modern use of psychotropic drugs that effectively alter individual brains' biochemical configuration. In 1994 Peter Kramer discussed the advent and the successes of Prozac, a mass-marketed selective serotonin re-uptake inhibitor for subjects in the clutch of melancholy (see also Lawlor 2012; Solomon 2002). Kramer deals with concrete North American culture and he circumnavigates philosophical questions effectively but frequently alluding to political issues. *Dramatis persona* is always the singular patient with a unique anamnesis. Kramer finds that these drugs have direct results on an individuals' social existence. "Serotonin levels influence social status" and that in animal testing "the results were dramatic" since "whichever monkey got Prozac, or a drug with a similar effect, dominated" (1997: 213–14). Brain-formation is continued into group-formation. Neural tissue becomes something rather uncanny in Kramer's ongoing assessments. Like Rose, Kramer follows the individualist agenda but with *Listening to Prozac* he indicates that traditional *humanist* ideas of maturity, prowess, and responsibility might have to be revised. Science and the humanities, the market value and the individual assessment of self-value, enter a relationship of mutual dependency with synaptic tissue as the currency in perpetual exchange. This focus on the nervous system gives way to rhetorics of technocracy. Kramer points out that "we are entering an era in which medication can be used to enhance the functioning of the normal mind" and *not only* the sick or afflicted (1997: 247). The notions of *enhancement* and *functioning* remain largely unquestioned for long passages of his book but he asks for a "Message in the Capsule" in his last chapter nevertheless. The line between self- and health care becomes political: "There is always a Prozac-taking hyperthymic waiting to do your job, so, if you want to compete, you had better take Prozac, too" (1997: 273). One better masters one's brain to master the world. Palahniuk's zombie kids found a deviant denouement to this task. The modern brain is a project, a responsibility, a part of one's CV, and an asset. This is a continuation of the solipsism Rose associates with a human's material, *synapto-chemical* existence.

Kramer folds plasticity up to a social level—a new tortuous material culture emerges at the end of the twentieth century in which the brain is weighed and amplified in economical exchange. An unapologetic perspective is helpful here to surmise the beginnings or continuation of posthumanism, a concept of culture and scholarly perspective that does not take the individual as the primary frame of reference. Despite the dystopian qualities of Kramer's pharmacological account, this must not be understood as *in*human or *anti*human—however, there is a macroscopic dimension to Western brains that has the capacity to alter the microscopic and individual potentials of plasticity. The brain is in-between—it is both an object and a subject of neuronal change that takes place in a diminutive synaptic area. One is confronted with almost inconsiderable connections and dependencies, the one alluded to here is between shareholder values on the one side and synaptic tissue in the specific brain region of singular citizens on the other.[4]

How could this be conceptualized? Simple neuroplasticity that is limited to the human skull is too simple a concept to include this new complexity. Traditional principles such as nature|nurture seem exhausted as one tries to overcome the neglect of "the body and the world beyond" (Pepperell 2003: 13) In *Neuropolitics: Thinking,*

*Culture, Speed* William Connolly deals with the brain's expanse. He asks for the entanglement of what is called a brain with what is called the collective|culture|outside: "How radical is the difference between concentrating your mind and taking Prozac to clear it of depressive thoughts?" (2002: 103). Depressive thoughts do have a chemical aspect but a general definition of them remains unobtainable. The individualism and solipsism Stephen Rose underlined support the inimitability of individually molded neural structures—the suffering or obstruction of thought is always as unique as the sufferer or the obstructed. Connolly complicates this: "Does the presumption that the difference is one of kind rather than degree tacitly invoke a conception of the supersensible itself open to contestation?" (2002: 103). How much discretion and control could the individual obtain here? She is rather a result of circumstances that *include* the brain as they include pillboxes in the bathroom cabinet. With the "supersensible" he relates to the Kantian notion of *noumena*—the things-themselves that cannot be grasped by humans who can only navigate in the material and mundane world of *phainomena*. This dualism (a rich and popular part of the humanist agenda) might crack as the modern world is confronted with the baffling object that is the brain. Connolly offers a nondualist notion and strengthens a new materialism that relies on a revaluation of the mundane, impersonal world which was frequently spurned in modern philosophy:

> Thinking is not merely involved in knowing, explaining, representing, evaluating, and judging. Subsisting within these activities are the inventive and compositional dimensions of thinking. To think is to move something. ... The cognitive—that is, the representational and explanatory—dimension of thinking coexists with its expressive, creative, and compositional functions. (2002: 104)

This notion of a nonlinear holism becomes crucial to any valid concept of the brain as it relies on plasticity and thus on matter persisting in and changing through time. The necessity to think *in time* and differentiation and less *in place* and set identities becomes apparent: acknowledging plasticity, the becoming of involved and involving neural matter, is a necessary step toward a sophisticated concept of cranial matter. This way the brain gets connected to grander, even more implicit and thus highly philosophical notions such as time, (re-)cognition, and communication. Connolly finds even more monumental notions as he discusses "complexity," "agency," and "vicissitudes" in his *A World of Becoming* (2011: 17, 43).

A circle closes: since melancholy was introduced as a somewhat *naked* philosophy before one can now acknowledge the magnitude and the ubiquity of mental processes that enter a no-man's-land of rather abstract categories. The thinker|melancholic finds himself or herself not in a world of humans and humanity but in a world of becoming. Here, the established human discourses based on language and communication cease to matter. Connolly eventually expands this into a discussion of spirituality and theology (2011: 97, 104). He states: "In a world of becoming, God and humanity are co-present" (2011: 107). Like melancholy, spiritual thought is wrestling with assessments of the whole and the limits of understanding. His particular notion of a nondualist (probably rather Eastern) spirituality cannot be understood as tranquility and calm that might somehow soothe the troubled mind. As the figure of *Melencolia I* sits and wonders, time passes but is also *minded*. Duration matters because there is no discernible movement or tendency toward salvation or redemption anywhere: the possibly religious points of reference (angelic figures, light rays on the horizon) do not help in Dürer's composition, at least they do not guide the thinking figure to a safe haven of knowledge. As the sole protagonist of the image she does not support an idea of humanity in charge and/or in control of things. She is *not* communicating or preparing to reach out. Melancholy can be understood here as thought that is declutched from humanity and the humanist project. It is also a move away from the progressive Kantian philosophy that privileges humanity at the threshold between the *phainomena* that are experience and the *noumena* that can only be referred to.

## Four: An object of affection in-between and nowhere

Classical humanities and cutting-edge science begin to overlap in the brain. Posthumanism calls for it and a study of melancholy demands it. Sometimes the brain remains a rather passive placeholder, an empty reference to an empirical authority. It becomes a reason or a mover, sometimes even a prime mover in a faux-Aristotelian sense. A neurophilosophy (a *melanchology* maybe) that is not in steady clinch with science is possible. One is to compile a rather wieldy but by no means reductionist concept of what a brain and/or neural tissue is (or rather *does*) while taking a look at *life itself* into account. The two cultures are to meet (Snow 2003). At least two fields of very recent philosophical research promise applicable results—both explore the in-between and a *situatedness* of things.

First off, one can consider the brain a site of *affect*. It is a key term to understand the brain as an evasive object of study that is rather *in-between* than in a fixed place. In the clinic, a healthy human has to be in a specific affective order (American Psychiatric Association 2013: 151).

Philosophy enables a different perspective since it uses *affect* in a much more modest manner. It still has to be handled with care though as it remains utterly contextual. *Affect theory* does not immediately apply terms like *function* and *causality* but actually disassembles them as traditional conceptualizations. In the *Affect Theory Reader*, Gregg and Seigworth underline the relevance of the body and the material world in this way of thinking as they write: "Affect arises in the midst of *in-between-ness*: in the capacity to act and to be acted upon [their emphasis]" (2010: 1). There is already a strong conceptual vicinity to plasticity here: both terms are rather hinging upon time than upon place. Actually, both affect and plasticity cannot be understood in the simple dualism of object and subject. The brain is fundamentally *in-between*. Kramer's and Connolly's remarks also led toward this problem of *locating* the brain—or the admission that it can neither be framed easily nor simply joined with notions such as personality or self. Gregg and Seigworth continue: "Affect is an impingement or extrusion" and "at its most anthropomorphic, [it] is the name we give to those forces—visceral forces beneath, alongside, or generally *other than* conscious knowing, vital forces insisting beyond emotion" (2010: 1).

Several terms here can easily be related to the condition of the naked philosopher, the brooding thinker. The melancholic is in a state of affection. There is an aspect of being "overwhelmed,"[5] which means that there is not a simple accessible exit or end to this situation for the subject. The brain's (or the entire nervous system's) activity involves more than conscious [and communicable] knowing—wrestling with "the world's apparent intractability" the melancholic puts the head on a fist and experiences the limits of cognition and the traction of consciousness (2010: 1). "Indeed, affect is persistent proof of a body's never less than ongoing *immersion* in and among the world's obstinacies and rhythms, its refusals as much as its invitations" (2010: 1). Immersion refers to a body enveloped in something bigger—the state of melancholy is immersive, just like consciousness and thought is rather *intensive* and total than neither *extensive* nor eventually finite. It denotes an end of neutrality as such: life continues in a steady en—and development of bodies.

The central brain is effectively dethroned in the philosophy of Deleuze and Guattari, which notoriously prefers rhizomatics, *in-betweens*, over hierarchical orders (Deleuze and Guattari 2004). Building upon Spinoza, affect and affection is to be understood here as "the ability to affect and be affected," as translator Brian Massumi writes in *A Thousand Plateaus*. "It is a *prepersonal* intensity corresponding to the passage from one experiential state of the body to another and implying an augmentation or diminution in that body's capacity to act" (2004: xvii). The term *affect* encourages scholars to open their perspectives to a grander concept of life.

With the notion of affect one can escape the humanist frame and actually grasp what the melancholic is wrapped up in: the inconceivability of thought as it occurs *with* but not *in* the brain. With affect, one can indeed conceptualize the brain as more than a legal tender or as an explanatory device or as a mere *thing*. Senses are involved, explicitly and implicitly, and the nervous system is not to be sketched by (human) cognition and emotion alone. Pharmacology and medicine perpetually reaffirm an ethics that necessarily puts the human being (as the patient|consumer) into the center of things. Here, the brain is inside the human, the synapses are inside the brain, and consecutively one is confronted with a matroshka-like concept of human life—it is both incompatible with the posthuman era Pepperell expects and the vast scope affect theory claims to relate to.

With affect, science and the humanities could begin to look for a common denominator, which would be, for the lack of a better word, *life*. A similar effort to merge the camps comes from Alva Noë. To any centralist concept of the brain he responds in *Out of Our Heads*: "You are not your brain. We are not locked up in a prison of our own ideas and sensations. The phenomenon of consciousness, like that of life itself, is a world-involving dynamic process. We are already at home in the environment. We are out of our heads" (2009: xiii). He defiantly opposes the notion of an inside with the outside in an interdisciplinary vein. Asking for the end or limits of a self is not really possible for him: "Our lives take place in a setting" (2009: 75). Any mechanistic narration surrounding Prozac and other drugs becomes less feasible (2009: xii).[6] Plasticity has finally left the skull: as a concept, it now applies to an entire world, a "larger system" that is in processes of becoming and that yields emerging consciousness. Affect theory urges scholars to be aware of reductionist and ordering habits of analysis that, in the case of the brain, may do nothing but obstruct the view.

Noe emphasizes the you, the discrete human being in his book. Eugene Thacker, however, asks for a much more comprehensive approach to existence than biology (the field of study presumably closest to the brain) can enable: "What if life is not reducible to biology—but also not reducible to 'consciousness', 'spirit', or 'intellect'? What if life is never self-evident in lived experience?" (2010: x).[7] Thacker's vantage point is cosmic and all-encompassing and belittles humanism even more. This questioning for the grander scheme of things has deep connections to the naked philosophy described above. For Steven

Rose, "The brain is biology's greatest challenge" (1978: 21). One could counter now that a theory for life as a whole is a grander challenge in which all academic disciplines are asked to contribute. Plasticity has to be adjusted as a concept. Claire Colebrook reckons that a *postbiological* brain as a new academic controversy promises the end of the human intellect's centrality:

> The brain's plasticity can at once serve as a model for a life that has no privileged models, at the same time as the brain needs to be demoted from a privileged position of "mind." The brain is a figure or image that intervenes in the history of ideas to enable a rethinking of the very notion of an idea ... one could see meaning as the property of living systems, from which the brain would be a sophisticated derivation. (2010: 29)

Melancholy is attached to this development: as a crisis of meaning, it could even be magnified as a crisis of life, incarnating itself in individual brains and leading to manifold ways of self-destruction or zombifications.

The notion of a constantly resurfacing and readjusting (thereby *living*) thought is continued by Robert Pepperell in his *The Posthuman Condition*. He locates consciousness explicitly *beyond* the brain. Thought is here part of the world, not a detached activity. He presents an *"energetic theory of mind* in which human thought, meaning and memory is understood in terms of the activity of an energy regulating system" (Pepperell 2003: 100). In concord with affect theory,

> the human is in essence no different from any other such "energistic" system we may find in the universe. ... If we can start to see how the most "sacred" of human attributes [including decision-making and self-care] ... operate in ways not dissected from other functions in the universe, then we are *moving away from the notion of humans as unique,* isolated entities and towards a conception of existence in which the human is totally integrated with the world in all its manifestations, including nature, technology, and other beings. (Pepperell 2003: 100)[8]

Not only does affect theory encourage research that examines the un-centered relationship between consciousness, the universe, and material conditions, but also it shares its skepticism with object-oriented ontology against the Kantian, humanist, anthropocentric framing of the brain. Quentin Meillassoux calls this Western tradition "correlationalism" or philosophies of access. For him, "It is possible to consider the realms of subjectivity and objectivity independently of one another" (2008: 5). He pleads, similar to Thacker, for a "naive realism" that does not favor consciousness and/or the human horizon of the world (2008: 3, 5). Neuroscience becomes conspicuous as a possible heir or squire of correlationism, of Western thought's persistent interest to put the sensory, individual and above all *human* existence into relation to a world it can experience. Neuroscience is a project of finding access to how this species "ticks" and is looking for access to the vexing unknown territory that is the brain.

Affect theory reminds scholars to pay attention to the billowing "bloom-space of an ever-processual materiality," one in which the melancholic state unfolds and wrestles with (Gregg and Seigworth 2010: 9). From a different vantage point, object-oriented ontologies help to consider a world that does not depend on human beings, a notion of life that is not based in its witnesses and their consciousnesses (Thacker 2010: x). According to Graham Harman, objects constantly withdraw from human focus as they "[are] existing in their own right ... the reality of objects is never fully deployed in their relations" and that this philosophy "holds that the human-world relation has no privilege at all" (2010: 69, 119). The ruminating melancholic actually is on this quest for a real object to grasp, to make sense of the relations of the world in order to identify a certain order. She is across with the usual "human-world relation."

This perspective can also be made especially explicit by the ambiguous notion of the brain "itself": it perpetually refolds through plasticity on the neurological level. One can address it with a noun but actually it is more than the sum of its parts. It is always deeper than the sum of relations it has to other objects: a brain relating to some other entity becomes a different object. For example, the perception of a tree changes a brain into a new entity which is *a tree-perceiving brain*: "If my relation with a tree forms a new object, then I as a real piece of that object find myself on its interior, confronting the mere image of the other piece" (2010: 117). There is no clear, substantial, nonchanging anchor that opposes the world. Immersion and embeddedness matter here instead of *phainomena* relating to *noumena*.

It seems that, confronted with the enigma of the brain, one is urged to conceptualize with diligence. In order to come up with a hefty "brick" that can be thrown or stacked, a doughty break of the privilege of unity, identity, and enduring definitions becomes crucial (Massumi 2004: xiii). Up until here, science and philosophy were uncomfortable travelers squeezed onto a bike unable to steer and pedal harmoniously. The travels can continue but only if two things happen. First, both camps are to find a common ground with basic notions such as affects and objects, energy and life. Secondly, one more passenger must be allowed to hop onto the already groaning vehicle. This passenger is *art*, the field in which a phenomenon like melancholy is expressed without justification, in which it is posed as a problem, as an enduring motif from before *Melencolia I* until today. In order to understand a minoritarian struggling consciousness, how it is communicated and expressed and urges to be dealt with in a noncondescending way, the Arts must relate to philosophical and scientific reflections.

## Five: The in-between on the screen

Asking what (and where) a brain *is* becomes asking what a brain *does*—at least when one considers how fatal and brutal the brain's products, thought and consciousness, can strike down. The one art that appears highly appropriate to open this horizon is cinema. "Attention to cinema can … inform these explorations" and "contemporary cinema techniques … heighten our powers of perception; alert us to complex relays among affect, thinking, technique, and ethics; teach us how to apply pertinent techniques to ourselves; and reveal things about the constitution of time that might otherwise remain hidden" (Connolly 2002: xiii, 2). And for Alain Badiou,

> A film is a proposition in thought, a movement of thought, a thought connected, so to speak, to its artistic disposition. How does this thought exist and get transmitted? It's transmitted through the experience of viewing the film, through its movement. It's not what's said in the film, it's not how the plot is organized that count [*sic*]; it's the very movement that transmits the film's thought. (2013: 18)

Depression or melancholy, this vexing crisis of meaning and centeredness, can be explored in this peculiar medium that deals with movement, time, and the montage of worlds—here, stagnation and rest become crucial, observable states. Film does not show *the* (fixed) world but opens *a* world and thereby *thinks*. Deleuze writes in *Cinema 1*: the shot "continuously divides duration into subdurations which are themselves heterogenous, and reunites these into a duration which is immanent to the whole of the universe" (Deleuze 1986: 20). Here is the flickering of thought and the movement of rumination: "We can say of the shot that it acts like a consciousness. But the sole cinematographic consciousness is not us, the spectator, nor the hero; it is the camera [which is only] sometimes human" (Deleuze 1986: 20). Melancholy is all-encompassing thought, it deals with parts and wholes and their relations. New objects form steadily, new configuration emerge perpetually: "The shot, that is to say consciousness, traces a movement which means that the things between which it arises are continuously reuniting into a whole, and the whole is continuously dividing between things" (Deleuze 1986: 20). This is what reflective thinking is and what a naked, intimate philosophy consists of. *Melencolia I* is declarative, film can be explorative—like thought. Cinema has the potential to surmount the assumed centrality of the brain, because here "the brain is nothing but this—an interval, a gap between an action and a reaction. The brain is certainly not a centre of images from which one

could begin, but itself constitutes one special image among the others"
(Deleuze 1986: 62). Rather than positing the brain as a fixed *container*
of thought, films are constructing wholes and do thereby justice to the
indomitability of thought.[9]

Films are minding the world on the basis of the physical textures and
occurrences life presents and allocates. Cinema is, unlike literature, a
medium that depends on matter and the light it emits. It is not its slave
but even develops fictitious matters: even recent CGI and animation
films are eager to deal with matters and materialities. Films are filled
with stuff, debris, and objects. Matter is the medium as these objects,
in the vein of Harman's theorems, change their roles and shapes as
they drift into and out of relevance. Cinema can make the depth of
objects palpable as they are parts of a dynamic composition—it can
lead to thinking and to ruminate like the melancholic does as the
camera-consciousness lingers, explores, and composes.

Lorenz Engell sees the movie theater as a place one can go to in order
to *see thinking*, to witness thinking *from an outside* and *to be thought*
oneself (2010: 137). Films do not show the way out of something, film
does not offer solutions in a plain copy of the supposedly real world
*out there*—seeing them as plain escapism and entertainment does not
do them justice. Films can be companions of the melancholic; they can
assist and provoke the ruminating mind—as they invoke questions
regarding the separation, the minding of time, duration, and movement.
Film is inherently melancholic for Robert Arnheim who writes in 1963
that film is not destined to confirm an outer reality. It does not reassure
the viewer by exuding a certain truth or *showing* something. To the
contrary, it is the medium to behold the *immeasurable*, *uncertain*, and
*unfathomable* world, thereby rather questioning the common forms
and formations of reality (Arnheim 2004: 405).

An unthinkable and insurmountable future led Palahniuk's heroes to
desperate measures. They adhered to the current revised phrenologies,
the new cults surrounding the plastic brain. What they actually needed
was probably just to sit down like *Melencolia* and think relentlessly
with a clenched fist. Any movie theater has many seats to choose from.

"Film awakens as much as it enfolds you," writes Stanley Cavell (1979: 17). It may also be a cultural technology that offers succor against the tyranny of a ferocious consciousness for all ages.

# Notes

1  The figure is thought to be female here; however, it is possible to question its gender and of course melancholy and depression are unisex.

2  Echoing C. P. Snow who identifies two coexisting "cultures" of academic questioning which are inapt to share resources until now. "Between the two [is] a gulf of mutual incomprehension—sometimes (particularly among the young) hostility and dislike, but most of all lack of understanding" (Snow 2003: 169).

3  In this point, melancholy shares aspects with fear, hysteria, or paranoia.

4  Plasticity blurs into the economic adaptability of a company with branded products and marketing techniques on the one side and into the civilian survival of consumers|employees with their own brain-care on the other. This signifies a nonlinear perpetual synchronization of brain matter, yielding a mass culture that is autocorrective to an unprecedented degree.

5  In contrast to being *normal*-whelmed like the majority of regular citizens.

6  The question of immersion and affective intensities matters here because Noe is emphasizing the category of space, surroundings, and proximity: "To understand consciousness—the fact that we think and feel and that a world shows up for us—we need to look at a larger system of which the brain is only one element" (Noe 2009: 10). With Noe one can consider the brain as an entity of *wide affect*. This embedded mind cannot be localized within a collective or within a heap of cells and is not a passive *or* active agent in certain surrounding but both.

7  He continues:

The very concept of life itself begins to dissolve and dissipate, while still remaining in use and in circulation. What if life is not assumed to reach its pinnacle in human life? What if life is only incidentally, and not fundamentally, an anthropocentric phenomenon? And what if life actually has very little to do with the presumed self-evident nature of the living? (Pepperell 2003: x)

8   The posthuman impetus is not a misanthropic one: it is merely the modest avowal that this species is not at the center of things, neither spiritually nor evolutionary or any other way. This is compatible with the depiction of melancholy on Dürer's image: the figure's attention is led toward some other place—which is not in the frame of the image. There is no authority she adheres to and the gaze remains uncentered and indistinct. The figure is in the middle of things with multiple objects and tools that can be tried out, applied, made useful—but nothing is used to a certain effect. The figure is somewhat occupied, she is busy—but nothing gets built or deciphered or done. There is energy at work since the figure is pressured by it to think—although it is not used to build something with the things at hand.

9   "The whole is … like thread which traverses sets and gives each one the possibility, which is necessarily realised, of communicating with another, to infinity. Thus the whole is the Open, and relates back to time or even to spirit rather than to content and to space" (Deleuze 1986: 16–17).

# References

American Psychiatric Association (2013), *Diagnostic and Statistical Manual of Mental Health Disorders: DSM-5*, 5th edn, Washington, DC: American Psychiatric Publishing.

Arnheim, R. (2004), "Die ungeformte Melancholie," in H. H. Diederichs (ed.), *Geschichte der Filmtheorie*, 403–16, Frankfurt: Suhrkamp.

Badiou, A. (2013), *Cinema*, Cambridge: Polity.

Barthes, R. (2003), *Mythen des Alltags*, Frankfurt: Suhrkamp.

Burton, R. (1883 [1621]), *The Anatomy of Melancholy*, Philadelphia: Claxton. Available online: http://ia600406.us.archive.org/34/items/anatomyofmelanch00burt/anatomyofmelanch00burt.pdf (accessed November 10, 2014).

Cavell, S. (1979), *The World Viewed. Reflections on the Ontology of Film*, New York: Viking Press.

178        *Media\Matter*

Colebrook, C. (2010), *Deleuze and the Meaning of Life*, New York: Continuum.

Connolly, W. (2002), *Neuropolitics: Thinking, Culture, Speed*, Minneapolis and London: University of Minnesota Press.

Connolly, W. (2011), *A World of Becoming*, Durham and London: Duke University Press.

Deleuze, G. (1986), *Cinema 1: The Movement-Image*, trans. H. Tomlinson and B. Habberjam, Minneapolis: University of Minnesota Press.

Deleuze, G. (1994), *Difference and Repetition*, trans. P. Patton, New York: Columbia University Press.

Deleuze, G. and F. Guattari (2004), *A Thousand Plateaus—Capitalism and Schizophrenia*, trans. B. Massumi, New York: Continuum.

Engell, L. (2010), *Playtime. Münchner Filmvorlesungen*, Konstanz: UVK.

Gregg, M. and G. Seigworth (2010), *The Affect Theory Reader*, Durham: Duke University Press.

Harman, G. (2010), *The Quadruple Object*, Washington: Zero Books.

Hebb, D. (1949), *The Organization of Behavior*, New York: Wiley & Sons.

Jamison, K. R. (2000), *Night Falls Fast: Understanding Suicide*, New York: Vintage.

Klibansky, R., E. Panowski, and F. Saxl (1992 [1964]), *Saturn und Melancholie. Studien zur Geschichte der Religion und der Kunst*, Frankfurt am Main: Suhrkamp.

Kramer, P. D. (1997), *Listening to Prozac*, London: Penguin.

Lawlor, C. (2012), *From Melancholia to Prozac. A History of Depression*, New York: Oxford University Press.

Massumi, B. (2004), "Notes on the Translation and Acknowledgements," in G. Deleuze and F. Guattari, *A Thousand Plateaus—Capitalism and Schizophrenia*, xvi–xix, New York: Continuum.

Meillassoux, Q. (2008), *After Finitude. An Essay on the Necessity of Contingency*, New York: Continuum.

Noe, A. (2009), *Out of Our Heads*, New York: Hill and Wang.

Palahniuk, C. (2013), "Zombie," Available online: http://chuckpalahniuk.net/news/zombie-a-new-original-short-story-by-chuck-palahniuk (accessed November 10, 2014).

Pepperell, R. (2003), *The Posthuman Condition. Consciousness Beyond the Brain*, Portland: Intellect Books.

Rose, S. (1978), *The Conscious Brain. Revised Edition*, New York: Penguin.

Sacks, O. (2010), *The Mind's Eye*, New York: Knopf.

Schwartz, J. and S. Begley (2003), *The Mind and the Brain: Neuroplasticity and the Power of Mental Force*, New York: HarperCollins.

Snow, C. P. (2003), "The Two Cultures," *Leonardo*, 23 (2 and 3): 169–73. Available online: http://www.jstor.org/stable/1578601 (accessed November 10, 2014).

Solomon, A. (2002), *Noonday Demon. An Atlas of Depression*, New York: Touchstone.

Sturma, D., ed. (2006), *Philosophie und Neurowissenschaften*, Frankfurt am Main: Suhrkamp.

Thacker, E. (2010), *After life*, Chicago: University of Chicago Press.

# 9

# The *Media Boundary Objects Concept*: Theorizing Film and Media

Florian Hoof

This chapter critically explores the recent trends to include theoretical frameworks from science and technology studies to film and media theory. In this context, the concept of material semiotics, an approach that enhances the semiotic model from the domain of signs to material objects and technology, is introduced to compensate for the existing shortcomings in film and media theory. Material semiotics is perceived as a feasible option to overcome the existing dichotomy in film and media theory, which is characterized by approaches that are either based on the concept of representation or centered on a strong notion of technology. Although, at first glance, these theory imports solve the issue of technological and image-as-text centrism, they also introduce the problem of relational determinism into film and media theory. Drawing on the existing criticism concerning the relational determinism of material semiotics as a starting point, I propose to theorize media from the perspective of the social worlds framework. I argue for the concept of a historical media epistemology based on the boundary objects approach developed by Susan L. Star and John Griesemer. I extend the idea of boundary objects to the concept of media boundary objects as a means of conducting historical analysis on media. The media boundary objects concept I propose here is a basic theoretical framework for historical as well as systematical research in media and culture. From this perspective, film and media are not limited to provide for the possibilities for communication or to increase the connectivity in a social system. Instead, they can also be conceptualized as structures that stabilize differences and nonsignificant boundaries in society between entities that do not communicate.

## Fuzzy material: Semiotic and technological determinism in theorizing media

This first image (Figure 9.1) shows material in its most basic form visible to human spectators. These are molecules in motion, known as the Brownian motion named after Robert Brown, a British botanist. In 1827 he was studying under a microscope pollen grain suspended in water. Here he observed a jittering motion: material seems to be in motion. In 1905 Albert Einstein drew on the Brownian motion to falsify the classical approach of thermodynamics. Objects could no longer be described as material that was moved by external force. The objects themselves, the material was in motion. This blurred the line between the idea of a distinct and stable object and a separate world outside the materiality of the object that was in motion. But neither Robert Brown nor Albert Einstein, who only provided a mathematical equation of the Brownian motion, could prove the existence of this motion. Three years later in 1908, Jean Baptiste Perrin, a French physicist, was finally able to prove the validity of the concept of the Brownian motion. He connected a microscope to a cinematograph. As a result he was able to verify the existence of the Brownian motion by filming these jittering particles.

At first sight, one might argue that this scientific film provides a representation of the Brownian motion. On closer inspection, however, it becomes obvious that the film does not show the Brownian motion, but a bunch of irregularly jumping dots. There is a simple reason for that: the motions of the molecules are too fast to be captured by the film camera. Therefore, others have argued that the Brownian motion was not verified by the visual representation of the movement in the single frames of the film, but through the technological apparatus of

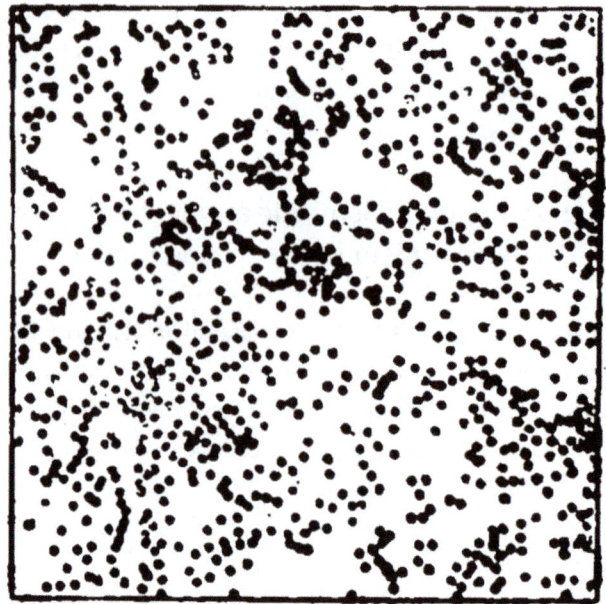

**Figure 9.1** *Brownian motion*. Perrin, Jean. 1910. "Die Brown'sche Bewegung und die wahre Existenz der Moleküle." Kolloidchemische Beihefte Band 1 6-7: 254.

film (Curtis 2005). The technological framing of the film that shows jumping dots stabilizes the assumption that these particles actually move, although the movement is invisible. It is neither the visual representation nor the mechanical transportation of film stock in the camera that finally enabled Jean Baptiste Perrin to validate the theory of Albert Einstein and the existence of the Brownian motion, but rather the interplay, the combination of film technology and the filmic image.

This small example sheds light on circumstances that can be described as problems of theory. How to analyze the somehow fuzzy interplay between the visual and the technological sphere? The initial distinction between the visual and the technological sphere can be traced back to a dichotomy in film and media theory. On the one hand, scholars focus on the dimension of visual representation and the filmic image; on the other hand, they focus on technology and infrastructure that enables, reproduces, transmits, and stores these very representations. In this context, the Brownian motion serves as a paradigmatic example to identify and to show the implications of this epistemological structure in film and media theory.

Plato's allegory of the cave perfectly illustrates this dichotomy. He describes a situation where a group of people are chained to a wall of a cave. They watch shadows projected on the opposite wall by things passing in front of a fire behind them. Due to the fact that they cannot see the technology or apparatus that produces these shadows they begin to take these shadows for real. From their position they cannot identify the source of their reality. It is assumed that in case they were able to spot the technological apparatus, their concept of reality would be shattered. Consequently, the observers either focus on the moving images projected to the wall or focus on the technology that produces these expressions. In this model there is no analytical position that would allow for focusing on both elements at the same time. This dichotomy also clearly manifests itself in two broad strands in film and media theory. On the one hand scholars focus on the technological apparatus, on the other hand they focus on the image, the shadows on the blank wall.

Under the latter, one can gather paradigms of representation that define and analyze media-as-text. One crucial condition to make these semiotic approaches work is to separate the text as a map of meaning from its technological surroundings. Technology is reduced to a framing instance that enables specific semiotic processes, but it is defined as a relatively stable dimension that does not interact with the textual meaning, the content expressed by media. In the context of film theory, the "apparatus theory" showcases this methodological strategy in an exemplary way. Here, Baudry and Williams (1974) differentiate between the invisible cinematic apparatus and the cinematic experience. The latter is paradoxically guaranteed through but not influenced by technology. The cinematic apparatus that, among others, includes technological items such as a film projector and a screen, albeit guaranteeing the black box cinema itself, becomes a black box for semiotic analysis. This is no coincidence but rather done on purpose. Christian Metz argues that

> the importance of making this distinction between the cinematic and the filmic fact lies in the fact that it allows us to restrict the meaning of the term "film" to a more manageable, specifiable signifying discourse, in contrast to "cinema" which as defined here, constitutes a larger complex (at whose center, however, three predominant dimensions may be distinguished: the technological, the economic, and the sociological). (Metz 1974: 12)

Therefore, film semiotics operates within the universe of signs (Eco 1970). In case of the allegory of the cave, film semiotics would focus on the shadows that appear on the blank wall. The technology that enables this black box situation disappears or is reduced to a functionalist one, taken for a granted surrounding. For the apparatus theory, it serves as a stable dispositive that produces ideological effects via the film screen.

In contrast, a technological perspective on film and media does not focus on the blank wall but on the apparatus that generates the shadow images in the first place. Here it is argued that the "medium is the message" (McLuhan 1964: 7–23). Consequently, the technological dimension of media is key to analyzing the relations between media and society (Ong 1987; Carpenter and McLuhan 1960; Kittler 1999). To really understand media, one has to focus on its technological aspects and not so much on issues of representation or the so-called media content. From their point of view, representations are just downstream effects of the technological, deep structure of media. Consequently, in the case of the allegory of the cave, they would focus on the technology, the fire, and the process of shadow making as part of the "media [that] determine our situation" (Kittler 1997: 28). "We can only ever know about people what the media are able to store and transmit. What counts, therefore, are not the messages or content … but only their circuit arrangements, those diagrams of observability in general" (Kittler 1997: 30). Thus, key to understanding society are the underlying systems of technological media, because the "content of any medium is always another medium" (McLuhan 1964: 8). From their perspective it is the technological structure of the cave and their forerunners that determine how humans are able to make sense of the world in the first place.

Obviously, both approaches have their theoretical shortcomings and benefits. While the technological perspective neglects the visual and semiotic dimension of media, the paradigm of representation ignores the media's technological dimensions. In the case of the allegory of the cave, this would restrict the perspective either to the shadows on the wall or to the fire, the source of the shadows. The same would happen if applied to the example of the Brownian motion. To be able to capture the epistemological status of the medium involved, it is rather advisable to focus on the interplay between technology and representation. Jumping dots makes sense only if one knows the details about the filmic technology involved that created these images in the first place. Therefore, a methodology is needed that enables us to cover this overarching framework between moving images and technology.

## Material semiotics: Material as relational effect

Not at all surprisingly, this theoretical dilemma has not gone unnoticed in film and media theory. Therefore, in the last couple of years, theoretical approaches from the field of science and technology studies became prominent with film and media scholars. For some, these approaches provide the missing link between media technological determinism and models of representation. Notably, the Actor-Network-Theory (ANT) (Latour 1987, 2007), as the most successful approach that stems from the methodology of the "sociology of translation" (Callon 1986), has become a new common ground for media, film, and cultural theorists to rethink modernist categories such as technology, human, or the image. Particularly, the concept of "material semiotics" (Law 2010: 176) has been attracting attention. Here, semiotic concepts that before were restricted to the dimension of signs are extended to include technological as well as material objects into their maps of meaning. In this context, the notion of materiality became prominent to replace existing terms that reproduce modernist dualisms such as human versus nonhuman or nature versus technology. Describing the relations between this French school of science and technology studies and the so-called materialist turn, Law stated that "matter matters" (Law 2010: 173) but only if it is detectable. Thus, "materiality is understood as a relational effect. Something becomes material because it makes a difference, because somehow or other it is detectable. … No relation of difference and detection. No relation at all." (Law 2010: 173). In this definition of materiality it cannot be reduced to an essential material core or be directly linked to concepts such as nature. Material is thoroughly perceived as a social construction and thus as a relational effect.

This perspective, that no longer distinguishes between human and nonhuman, was implicitly perceived as an opportunity to overcome the division between the cinematic and the filmic fact proposed by Metz as well as the technological determinist perspective that defines media merely as "circuit arrangements" (Kittler 1997: 28). The ANT was described as a promising methodology to reconnect the technological with the representational or semiotic model on materialist terms without having to abandon one of these two dimensions. Existing categories such as *object* or *technology* then could be marked as part of a modernist epistemology and subsequently could be reformulated in a flexible, relational system. This new system heavily relies on the concept of performativity to prove their interconnectedness. While classical semiotics as well as technological media approaches is based on a front-end decision to include or to exclude specific phenomena from their field of study, ANT relies on a more flexible action-based concept informed by constructivism to validate their actor-network.

## Relational determinism: Network as metaphor

The turn toward materialist semiotics in science studies has not been without criticism. Law, too, states that with the issue of materiality "there are complications" (Law 2010: 173) regarding how to define and analyze relations between human and nonhuman actors under a materialist perspective. There has been a controversy concerning the asymmetric approach to human and nonhuman actors (Latour 1993) between the sociology of social knowledge (Collins and Year 1992) and the sociology of translation (Callon and Latour 1992). But besides this fundamental debate concerning the philosophical paradigms at stake, there has also been criticism concerning the very methods and models employed by the ANT and their approach of materialist semiotics. One dimension was concerned with the "version of semiotics" (Lenoir 1994: 124) the ANT employs. Lenoir wonders if "we are not led in the end, kicking and screaming, back into old style realism" (Lenoir 1994: 126). As Latour "bursts through the sign barrier" (Lenoir 1994: 126) he gains the "ability to move from signs to things and back" (Akrich and Latour 1992: 259) but he achieved this by introducing a new ontology that consists of concepts such as actors, hybrids, or actants. In this context, Lenoir is concerned that "we are provided a map and potentially a set of taxa that specify certain types of actors and narratives, and with this we are back to the old ground of realism and representation" (Lenoir 1994: 126). He continues to trace this realist impulse to the underlying ideas of Julien Greimars and his structuralist grounding of semiotics. Greimars "relies on a minimal set of atomic meanings, 'nuclear' semes modeled after chemical elements, replete with isotopies and homotopies. ... Greimars's approach borders on a reductivism to biological deep structures" (Lenoir 1994: 129). Or to put it differently, even if a researcher relies on a model of an endless relational network to describe the world, he has to frame and determine his object of study in the first place. In the case of film semiotics, Metz described this quite frankly as a strategy to get a "more manageable" (Metz 1974: 12) object of study. For approaches influenced by ANT the question of boundary work, how and under which theoretical premises to scale a relational network and to emphasize specific parts of a network, seems unanswered. Perspectives that rely on a technological or a semiotic perspective specifically address this issue with their reductionist focus on technology or world-as-text. Reason for the enormous success of the asymmetric anthropology proposed by Latour, one might suspect, is the promise that with this modeling of the fait social, problems of determinacy and reductionism can be avoided. But as Lenoir shows, material semiotics implicitly relies on relational network models from biology as well as physics to overcome technological determinism and at the same time the restrictions of semiotics.

From a film and media perspective, this change toward a relational network model has some significant downsides. Communication and media technology has been associated with and described by network metaphors since the very beginning. The mathematical communication model by Shannon and Weaver defines communication as a purely relational situation between sender and receiver (Shannon 1948). Relational models and network metaphors have been taken up by management consulting as well as by organizational theory since the 1880s. In the last decades, organizational structures have been conceptualized by relying on atomistic models provided by the consulting industry (Hoof 2015: 9–17). Corporate hierarchies have been replaced by project-based structures that can be traced back to the logics of network models (Boltanski and Chiapello 2003). Besides being a handy tool for social theory, they are also representations of a purely modernist idea of how to structure the social in society. Consequently, there are methodological issues concerned. This diminishes the analytical potential of a relational approach to explain parts of society as a result of modernist epistemology while using a similar logic to describe modernity. Here, the blind spot of the relational network model becomes visible.

One might argue that categories and dualisms such as human versus nonhuman as well as technology versus nature are cornerstones of nineteenth-century modernity. I would argue that the same is true for the network metaphor concerning the twentieth century. Lynch argued that the ANT is somehow trapped in the dualism of text and context that emerged from the semiotic and finally representational model of thinking. Even if one "broadens the field of contextual relations so that it includes an unholy mélange of intertwined human and non-human 'actants', the analyst is still faced with the problem of specifying which actants are relevantly part of the network associated with the … phenomenon under investigation" (Lynch 1994: 146). Or to approach this dilemma in the notions of system theory, the relational network model fails to take into account its own role as an observer (Luhmann 1999), in this case the very structure of modeling the social as a relational network structure. While focusing on network relations the boundaries within the network, the nonrelational aspects are undertheorized. Therefore, as Rudy puts it: "Latours descriptive explanations [are] based on a 'strong theory of the recording frame'—methodological strictures [*sic*] focused on summing up associations between actants and connections between networks—rather than theorizing boundary practices of inclusion and exclusion, materialist abstraction, and quantitative change and qualitative difference" (Ruby 2005: 119).

## Boundary objects concept

Besides the sociology of translation and ANT with the "social worlds framework" (Clarke and Star 2008), there has been a second strand of science and technology scholars who refused the Greimarian atomistic explanation of society. It focuses on the permanent tension between the formal and the empirical, the local and the situated, and attempts to represent information across localities. "It is this tension itself that is underexplored and undertheorized. It is not just a set of interesting metaphysical observations. It can also become a pragmatic unit of analysis. How can something be simultaneously concrete and abstract? The same and yet different?" (Bowker and Star 2000: 291–2).

Due to the fact that media theory is precisely an endeavor that by definition borderlines between the concrete and the abstract, I will have a closer look at this second strand of science and technology studies and what it has to offer for film and media theory to avoid relational network determinism. Here I specifically focus on the benefits of the boundary objects concept.

Susan L. Star and James R. Griesemer developed the idea of boundary objects clearly as a critique of the relational network model of the so-called sociology of translation (Callon 1986). While the latter in large parts emerged from sociological fieldwork in scientific research laboratories (Latour 1987; Latour and Woolgar 1986), Star and Griesemer conducted a historical case study on the Museum of Vertebrate Zoology in Berkley (Star and Griesemer 1989). In contrast to the spatial and organizational structure of a locally centered laboratory, the scientists here depended on amateur collectors that were spread all over the country. Although these people were not particularly interested in science, they helped to describe, systemize, and collect items for research in the museum. According to Star and Griesemer, they were "from different social worlds" (Star and Griesemer 1989: 388), while the staff in scientific laboratories was specifically selected, trained, and regrouped for the purpose of scientific research. So from their perspective it was quite surprising that the Museum was able to work properly.

From the perspective of ANT, a laboratory could be described as a functioning network that consists of a number of nodes between different kinds of actors. Star and Griesemer mainly criticized the network metaphor used. Thereby they argued that network logics are too closely associated with the very idea and structure of scientific research. According to Star and Griesemer, this concept of network focuses too much on a central authority, a person who manages the complex network and in the end decides what will be included in and what eventually will be excluded from the network. Their critique of the network paradigm specifically addresses the concept of the so-called obligatory passage points (Star and Griesemer 1989: 389–91) employed by the sociology of translation. Obligatory passage points are defined as an instance of selection that has to be passed in order to become part of a network or to stay in the network. Here a selection between diverse potential options that exist in the laboratory takes place. Consequently, several prospective nodes of the network are rejected at these obligatory passage points. That is exactly where the relational determinism described above kicks in.

When applied to the specific situation of highly formalized environments such as scientific research laboratories, the ANT concept turns out to be a productive methodological tool. If applied to phenomena that are characterized by a more heterogeneous, nonformalized, and nondocumented structure, this situation changes. The network metaphor as a scientific heuristic is not sufficiently capable to track down nonsignificant parts involved in such circumstances. Dysfunctional parts literally vanish at the obligatory passage points, as they are no longer part of a functioning network of actors and things. While technological and semiotic approaches clearly articulate what they will leave aside, this process of vanishing-in-the-making is less visible in the logics of ANT. Consequently, with the network metaphor one can analyze epistemological movements, and one can identify and describe different nodes of a research network, human as well as nonhuman actors. But at the same time it makes it difficult to answer why one of these nodes was successfully established in the first place. Furthermore, there are no answers to questions such as "Who carries the cost of distribution, and what is the nature of the personal in network theory?" (Star 1991: 44). Are there any alternative ways in existence that were not successful and therefore did not become part of the actor-network? How to define nodes that are not centered on a common objective? What are the criteria of selection in such a case? And, finally, how to explain the stable conditions under which the network operates? This becomes an issue of methodology especially if instruments originating in science studies are used for media and cultural analysis.

To deal with these methodological issues, Star and Griesemer developed the boundary objects concept. This approach emphasizes the improbability of communication and cooperation in social behavior by taking into account the heterogeneous character of social actors and technology involved.

> Boundary objects are objects which are both plastic enough to adapt to local needs and the constraints of the several parties employing them, yet robust enough to maintain a common identity across sites.

They are weakly structured in common use, and become strongly structured in individual site use. These objects may be abstract or concrete. They have different meanings in different social worlds but their structure is common enough to more than one world to make them recognizable, a means of translation. (Star and Griesemer 1989: 393)

To illustrate the boundary objects concept, Star and Griesemer draw on a brief example. From their perspective, an archive can be perceived as a boundary object. The archive itself, the building or the collection is a rather sturdy structure. It collects and stores items according to a specific classification system that is difficult to move or change, at least in a short period of time. Nevertheless, one can access it in a very flexible way. The form of a discrete relational storage system enables a nonlinear mode of access. You can access one item without thinking or relating to all the other items in store. This possibility of individual access guarantees a low-level threshold, which is crucial for an object to become a boundary object. It organizes the boundaries between different spheres or systems of society—to use the notations of system theory. People might have different reasons to access the archive or they are from different social worlds but they can coexist in this boundary object without having to directly relate to each other. There are no obligatory passage points in a boundary object; instead, boundary objects provide a structure that enables incoherent individuals to coexist without even having to relate to each other. But, with the boundary objects concept, one can describe this situation as loosely interlinked correlating actions. So, in contrast to the theoretical focus of the ANT and its attempt to explain the world as an interconnected network of performative action, the concept of boundary objects focuses on the conditions that increase the chances of communication or cooperation. This perspective on processes of knowledge and communication enables us to get a grip on latency, on things and processes that might not be relevant for a specific process at that very moment. The concept helps to track down the epistemological character of things that are not significant at that moment but might become significant later on.

## The media boundary objects concept

The concept of boundary objects defined as the description of the possibilities of communication and cooperation bears some similarities with the conception of media as a basic structure of communication. Bowker and Star argue that their idea of boundary objects can be paralleled with the idea of media proposed by Marshall McLuhan. Drawing on the example of an electronic information system, they exemplify their basic principle of the boundary object, which is the linking of form and action.

> "The medium of an information system is not just wires and plugs, bits and bytes, but also conventions of representation, information both formal and empirical. A system becomes a system in design and use, not the one without the other. The medium is the message, certainly, and it is also the case that both are political creations." (Bowker and Star 2000: 292)

McLuhan's dictum that it is the "medium that shapes and controls the scale and form of human association and action" (McLuhan 1964: 9) is also true for the concept of boundary objects. Elsewhere, Bowker and Star define their concept as "communication medium" (Bowker and Star 2000: 398). But these are no single media that become relevant as "they deal in regimes and networks of boundary objects" (Bowker and Star 2000: 313).

The boundary objects concept discloses some similarities to approaches deployed by film history in research on early cinema and turn-of-the-century visual culture (Gunning 1989; Griffiths 2002). One can even draw a connection to the idea of the "Discourse Networks" (Kittler 1992). In the dimension of aesthetics, it bears resemblances to the concept Jacques Rancière has described as an "aesthetic regime" (Rancière 2006).

The concept of "media boundary objects" (Hoof 2011) I propose here is a basic theoretical framework for historic as well as systematic research in media and culture. It draws on Marshall McLuhan's definition of media as an extension of man, but at the same time it pays attention to structural aspects of media and relativizes his rather vague optimism and the generalization tendencies on media and technology. From this perspective, media are not limited to providing the possibilities for communication or to increasing the connectivity in a social system. Instead, they can also be conceptualized as structures that stabilize differences and nonsignificant boundaries in society between systems that do not communicate.

## Media boundary objects in the wild: Film and consultancy

I deployed this framework in a study on the history of the consulting industry and their strong ties with media technology (Hoof 2015). In my archival research on the history of the early consulting firms between 1880 and 1930, I discovered what can be described as a visual culture based on a system of technological devices of business consulting and management. The emerging consulting industry relied heavily on film, photography, and other graphical devices to legitimate their new profession and their new business model. That is, to sell expertise to industrial manufacturers. One rather successful strategy was to conduct film-based time and motion studies in the factories to boost productivity. The result of these film studies was transformed into highly abstract forms of visual knowledge (Figures 9.3–9.4).

Not only were the consultants able to actually show inefficiency on film but also the abstract aesthetics appealed to the customers of the consulting firms. The devices and the visual forms created by graphical devices, such as film, represented a newly emerging form of knowledge, the consulting knowledge—a form of expert knowledge that deals with problems of coordination and of process steering and is thus not visible but a rather abstract form. But it also appealed to the workers. In the factories, workers who became objects of time and motion studies felt like little film stars (Figure 9.2). Here, the existing link to popular film and the star system in the film industry lent the film the function of a media

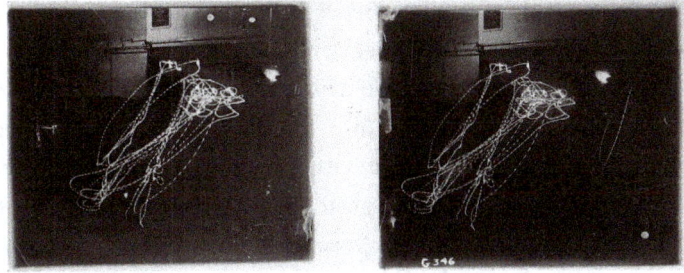

**Figure 9.4** *Stereoscopy*. TECHNOSEUM, Landesmuseum fuer Technik und Arbeit Mannheim, Nachlass Witte/Gilbreth, No. 1994-1241.

**Figure 9.3** *Filmstills*. Original Films of Frank B. Gilbreth. Prelinger Collection, Library of Congress.

**Figure 9.2** *Photograph*. TECHNOSEUM, Landesmuseum fuer Technik und Arbeit Mannheim, Nachlass Witte/Gilbreth, No. 2005-0769.

boundary object that mediated between the different socials worlds of the consultants, the factory owners, and the workers.

A common interpretation would be—taking a Foucauldian stance— that these are representations of a disciplinary discourse aimed at maximizing the workers' productivity. They could also be perceived as part of an aesthetic discourse of abstraction that originated in the field of arts at the end of the nineteenth century, as Rosalind Krauss argued (1979). But in analyzing the filmic representations and devices in their historical context, it soon becomes obvious that they were often not at all successful in achieving these objectives. In a great many of these cases, they failed to restructure companies despite being supported by these media devices. This raises the question of whether there are other epistemological dimensions attached to media-based consulting. Despite not being that successful, these different media devices—there is no doubt—mattered at that time.

I have argued that these instruments, technologies, images, and visualizations did not work in one narrow perspective; they functioned as media boundary objects that implicitly stabilized the system between the different social worlds involved. The effect of these media only becomes tangible if we look at the overlapping character of different kinds of medial aspects and not at the isolated categories of representation and technology. Conceptualizing film as a media boundary object enables us to analyze it as "something that has force to mediate subsequent action" (Bowker and Star 2000: 298). Because of its cutting-edge aesthetics, technology, and its connection to popular culture, it offered a low-level threshold. Everybody could participate in the media boundary object film because they were able to adopt or use it as a representation, as a technology, or as a social practice according to their own agenda.

Consequently, in the case of the consulting industry, the media employed can be described as boundary objects that "arise over time from durable cooperation among communities of practice" (Bowker and Star 2000: 297). According to this definition, film was at the same time weakly structured in common use, but strongly structured in individual site use.

Film was plastic enough to adapt to the local needs and constraints of the several parties employing it yet robust enough to maintain a common identity across sites. Here the media boundary objects approach enables a framework of historical media analysis while avoiding getting caught in representational, technological, or relational determinism.

# Conclusion

The "media boundary objects concept" I propose here is a basic theoretical framework for historical as well as systematical research in media and culture. From this perspective, film and media are not limited to provide for the possibilities for communication or to increase the connectivity in a social system. Neither are they reduced to a determinist perspective linked to an idea of technical media as deep structure of society. Instead, they can be conceptualized as structures that stabilize differences and nonsignificant boundaries in society between systems that do not communicate. The media boundary objects concept enables us to analyze film and media beyond the scope of technological determinism and image-centered models of representation without getting trapped in relational determinism. With this perspective in mind, let us return again to the film on the Brownian motion the article started with. While Curtis enlarged the perspective from the filmic picture to the technological apparatus, the concept of "media boundary objects" in film and media studies would be the next step toward reconceptualizing media by taking account of its social as well as epistemological implications.

# References

Akrich, Madeline, and Bruno Latour (1992), "A Summary of a Convenient Vocabulary for the Semiotics of Human and Nonhuman Assemblies," in Wiebe Bijker and John Law (eds), *Shaping Technology/Building Society: Studies in Sociotech nical Change*, 259–64, Cambridge, MA: The MIT Press.

Baudry, Jean-Louis and Alan Williams (1974), "Ideological Effects of the Basic Cine-matographic Apparatus," *Film Quarterly*, 2: 39–47.

Boltanski, Luc and Ève Chiapello (2003), *Der neue Geist des Kapitalismus*, Konstanz: UVK.

Bowker, Geoffrey C., and Susan L. Star (2000), *Sorting Things Out. Classification and its Consequences*, Cambridge, MA: The MIT Press.

Callon, Michael (1986), "Some Elements of a Sociology of Translation: Domestication of the Scallops and the Fishermen of St Brieuc Bay," in John Law (ed.), *Power, Action and Belief. A New Sociology of Knowledge?* 196–233, London: Routledge & Kegan Paul.

Callon, Michel and Bruno Latour (1992), "Don't Throw the Baby Out With the Bath School! A Reply to Collins and Yearley," in Andrew Pickering (ed.), *Science as Practice and Culture*, 343–68, Chicago: University of Chicago Press.

Carpenter, E., and Marshall McLuhan (1960), *Explorations in Communication*, Boston, MA: Beacon Press.

Clarke, Adele E., and Susan L. Star (2008), "The Social Worlds Framework: A Theory/Methods Package," in Edward J. Hackett, Olga Amsterdamska, Michael Lynch, and Judy Wajcman (eds), *Science and Technology Handbook*, 113–37, Cambridge, MA: The MIT Press.

Collins, H. M., and Steven Yearley (1992), "Epistemological Chicken," in Andrew Pickering (ed.), *Science as Practice and Culture*, 301–26, Chicago: Chicago University Press.

Curtis, Scott (2005), "Die kinematographische Methode. Das 'Bewegte Bild' und die Brownsche Bewegung," *Montage* AV, 2: 23–43.

Eco, Umberto (1970), "Articulations of the Cinematic Code," *Cinematics*, 1 (1): 590–605.

Griffiths, Alison (2002), *Wondrous Difference. Cinema, Anthropology & Turn-of-the-Century Visual Culture*, New York: Columbia University Press.

Gunning, Tom (1989), "An Aesthetic of Astonishment. Early Film and the (In) Credulous Spectator," *Art and Text*, 34: 114–33.

Hoof, Florian (2011), "Ist jetzt alles Netzwerk? Mediale 'Schwellen- und Grenzobjekte'," in Florian Hoof, Eva-Maria Jung, and Ulrich Salaschek (eds), *Jenseits des Labors. Transformationen von Wissen zwischen Entstehungs- und Anwendungskontext*, 45–62, Bielefeld: Transcript.

Hoof, Florian (2015), *Engel der Effizienz. Eine Mediengeschichte der Unternehmensberatung*, Konstanz: Konstanz University Press.

Kittler, Friedrich (1992), *Discourse Networks, 1800/1900*, Stanford, CA: Stanford University Press.

Kittler, Friedrich (1997), *Literature, Media, Information Systems: Essays*, Amsterdam: OPA.

Kittler, Friedrich (1999), *Gramophone, Film, Typewriter*, Stanford, CA: Stanford University Press.

Krauss, Rosalind (1979), "Grids," *October*, 9: 50–64.

Latour, Bruno (1987), *Science in Action. How to follow Scientists and Engineers through Society*, Cambridge, MA: Harvard University Press.

Latour, Bruno (1993), *We have Never been Modern*, Cambridge, MA: Harvard University Press.

Latour, Bruno (2007), *Reassembling the Social. An Introduction to Actor-Network-Theory*, Oxford: Oxford University Press.

Latour, Bruno and Steve Woolgar (1986), *Laboratory Life: The Construction of Scientific Facts*, Princeton, NJ: Princeton University Press.

Law, John (2010), "The Materials of STS," in Dan Hicks and Mary C. Beaudry (eds), *The Oxford Handbook of Material Cultural Studies*, 173–88, Oxford: Oxford University Press.

Lenoir, Timothy (1994), "Was that Last Turn a Right Turn?," *Configurations*, 2: 119–36.

Luhmann, Niklas (1999), *Die Kunst der Gesellschaft*, Frankfurt a. M.: Suhrkamp.

Lynch, Michael (1994), "Representation is Overrated: Some Critical Remarks about the Use of the Concept of Representation in Science Studies," *Configurations*, 2: 137–49.

McLuhan, Marshall (1964), *Understanding Media. The Extensions of Man*, New York: McGraw-Hill.

Metz, Christian (1974), *Film Language. A Semiotics of the Cinema*, Chicago: University of Chicago Press.

Ong, Walter J. (1987), *Orality and Literacy. The Technologizing of the Word*, London: Methuen.

Perrin, Jean (1910), "Die Brown'sche Bewegung und die wahre Existenz der Moleküle," *Kolloidchemische Beihefte* Band, 1 (6–7): 221–300.

Rancière, Jacques (2006), *The Politics of Aesthetics: The Distribution of the Sensible*. New York: Continuum.

Ruby, Alan P. (2005), "On ANT and Relational Materialisms," *Capitalism Nature Socialism*, 4: 109–25.

Shannon, Claude E. (1948), "A Mathematical Theory of Communication,"
    *The Bell System Technical Journal*, 7, 10: 379–423; 623–56.
Star, Susan L., and James R. Griesemer (1989), "Institutional Ecology,
    'Translations' and Boundary Objects. Amateurs and Professionals in
    Berkeley's Museum of Vertebrate Zoology, 1907-39," *Social Studies of
    Science*, 19: 387–420.
Star, Susan L. (1991), "Power, Technology and the Phenomenology of
    Conventions: On Being Allergic to Onions," in John Law (ed.), *A Sociology
    of Monsters. Essays on Power, Technology and Domination*, 26–56, London:
    Routledge.

Part Four

# [Art|Matter]

# 10

# Borderline: Nauman's Balls and Acconci's Shoot

Eva Ehninger

In *Understanding Media. The Extensions of Man* Marshall McLuhan utilized the myth of love-struck Narcissus yearning for his own image mirrored on the water surface in order to describe the double effect of media—in this case a mirrored image—on the human subject. According to McLuhan, any medium functions as an extension of ourselves and simultaneously serves as an auto amputation. Both effects of the use of a medium irritate the subject, who feels overextended and constricted at the same time. McLuhan ([1964] 2003: 64) explicates this claim by describing the wheel as an extension of the foot. Its successful implementation was a result of the physical and sensual acceleration of exchange by written and monetary media. But as much as the wheel served as a counter-irritant to the increased burdens of modern life, it brought about a new and equally upsetting intensity of action by isolating the function of movement from the physical activity of walking.

This kind of irritation, says McLuhan, needs to be relieved and numbed by the medium itself: "This is the sense of the Narcissus myth. The young man's image is a self-amputation or extension induced by irritating pressures. As counter-irritant, the image produces a generalized numbness or shock that declines recognition" (McLuhan [1964] 2003: 64). Due to that very image Narcissus, set outside himself, reaches an equilibrium and becomes numb, a closed system. According to McLuhan, this double effect has two consequences for humanity's accelerated use of and dependence on technology and its media. First, every new invention or technology demands new equilibriums and, second, in order to reach these equilibriums, we have to embrace technologies: "... We must, to use them at all, serve these objects, these extensions of ourselves, as gods or minor religions" (McLuhan [1964] 2003: 67–8).

In 1976, for her discussion of video as an artistic medium, Rosalind Krauss referred to narcissism as the psychological disturbance that borrows its name from Narcissus' myth. According to Krauss (1976: 50), the video monitor itself functions as a mirror: "... Narcissism [is] so endemic to works of video that I find myself wanting to generalize it as *the* condition of the entire genre." Krauss thus sets video apart from the other visual arts, since it is not so much constituted by formal characteristics but rather by the psychological condition of narcissism. This psychological state of the self, split (amputation) and doubled (extension) by its mirror reflection, constitutes for Krauss not the subject matter but rather the form, or model, of video.[1]

It is likely that McLuhan's general alignment of man's (or woman's) use of media with Narcissus' experience has fed into Krauss's assessment of the specific medium of video. There are some crucial points of comparison between the work of the media analyst and that of the art theorist that will prove important for my discussion of American body art and its modes of perception. McLuhan and Krauss focus their accounts on the human body. Media affect the body, and the subject needs to respond psychologically to this physical irritation. Also, both authors react with their texts to the phenomenon of "mass media," a recent development of the late 1960s and early 1970s. Whereas McLuhan responds optimistically to this increase in media by introducing a critical vocabulary and initiating a scholarly discourse, Krauss evaluates video and its narcissism as a symptom of an aesthetic world that has been "deeply and disasterously [*sic*] affected by its relation to mass-media" (Krauss 1976: 59).[2] For Krauss, video, by means of its narcissistic characteristics, both answers to and critiques the increased mediation of aesthetic experience. Video artists' interest in (their own) bodies as the subject and object of their work is in her eyes evidence of this critical attitude toward mass media. The control image on the monitor acts as a point of reference for the examination of the self-image and its archetypes, which are culturally and historically predetermined. In a video the human body as the most manifest residual of the real becomes a codification of contemporary society's yearning for an authentic experience.[3]

We need to keep in mind, however, that the codification of the human body begins at an earlier stage than through the video camera's monitor. The body itself in its material reality acts as a medium. Its video-documentation functions as a counter-medium which is in the process just as much revealed in its materiality as the human body is in its mediality. This argument is of importance, because it sidesteps the regular art historical division between (real, material) body and (artificial) medium, as well as the devaluation of a mediatized artistic experience, which even Krauss in her reassessment of video cannot shake off completely. Feminist discourse has theorized early and decisively about the codification of the human body.[4] Video has played an important role in this discourse as it was employed by artists and understood by critics to express and control individual body images that ran counter to culturally or historically determined stereotypes. Just as the presumed neutrality of the recording device is questioned so are the seemingly unencoded bodies it presents (Gever 1990; Spielmann 2005: 243–54).

While I agree with the authors' dismissal of the normative concept of an uninscribed body, the feminist perspective in my opinion has led to a deceptive semantic alignment of body art, the abject, and feminist theory.[5] The theoretical basis for the association of the "formless" with the "abject" may be found in Julia Kristeva's theory of the abject. Kristeva's aim was to describe the connection between subject and object, her interest in the "abject" was thus based less on social than on psychoanalytic and philosophical questions. In *The Powers of Horror* (Kristeva [1980] 1982), she discusses the abject as the arrested passage from subject to object for which the psychiatric term "borderline" would prove to be useful. Because the stoppage in this intermediary position is according to Kristeva due to an inability of the child to separate itself from its mother—it is caught up within a suffocating, clinging maternal lining, the mucous surroundings of bodily substances—the abject is regularly short-circuited with the female body, and its theory is applied to body art via a feminist reading of these artistic positions. Kristeva's conviction that there is no fundamental discontinuity between the production of a work of art and the life of the individual, and her continued interest in the artistic endeavor, which features prominently in her psychoanalytic writings, has aided this art historical line of argument (see, e.g., Lechte 2012).

An artwork that is irritating, even repulsive in its direct employment of the human body, seems to be worthy of critical discussion only if its offensiveness can be filtered directly into a theoretical discourse— if the *presentation* of the naked body, its private parts, their intimate openings, and offensive fluids is a *representation* of broader sociocultural discourses. According to Sabine Flach this shift can be understood as a historical development, with its starting point in the late 1960s. By then, she argues, the artists (or critics for that matter) no longer discuss a seemingly holistic body image, but instead use the body in aesthetic practices to illustrate those discourses that are being developed with and through the body (Flach 1994: 12).

American body art is regularly utilized as an illustration of postmodernist theories developed around the critique of modernist perception: first, criticism informed by psychoanalysis, which rejects the idea of a singular perspective that targets the object of perception from one dominant point of view. Hal Foster (1996: 113) defines this critique of the tradition of perception and the aggression against the visual with help of a wordplay: "'Obscene' does not mean 'against the scene', but it suggests an attack on the scene of representation, …" Second, and building on this input from psychoanalysis, feminist art criticism, which lambasts the male gaze and its one-dimensional focus on the female body-object.

The writings of Jacques Lacan served as a basis for both of these theoretical frameworks. Rosalind Krauss (1993) referred to him for her critique of the modernist notion of pure and homogeneous opticality, while Julia Kristeva built her theory of the objectification of women on a critique of Lacan's concepts (Kristeva [1969] 1980; Forte 1988: 220–1). A third critical perspective on modernist perception aimed at the institutionally determined dichotomy of artist versus artwork, and the institutional specification of the artwork as an "object" or "product" to be "consumed" by an "observer." Kirsi Peltomäki (2007) presents the viewing subject who is made aware of its individual position and perspective as a vital component of the institutional critique formulated in 1970s' performance, installation, or media art.

The material presence of a human body in these artistic practices seems to be acceptable only if it conveys such abstract concepts. The performances and video work of Bruce Nauman and Vito Acconci have regularly been applied to the argument of institutional critique, which in Nauman's case is regarded as more conceptual, in Acconci's case as more political. Both artists' focus on the mediative force of bodily matter—not its function as a signifier but its potential for individual signification via the physical parallelism of performer and observer— has up to now not been in the foreground of the discussion. My analysis of the mediality of human matter as presented in video will thus not only provide a modified perspective on these artists' practice. Based on the claim that the assessment of the cultural codification of the body does not have to begin and end with the issue of gender, it also aims to redirect the discussion about American body art. As a "cool medium," to use McLuhan's term ([1964] 2003: 39–50), the human body necessitates individual responses by each observer to fill in narrative and semantic gaps.[6] When mediatized by video, with its own semantic contexts, the body is not more but actually less definitively inscribed.

## Bruce Nauman's balls

For his film *Black Balls* (1969) Bruce Nauman positioned an industrial camera in extreme close-up in front of his body (Figure 10.1).

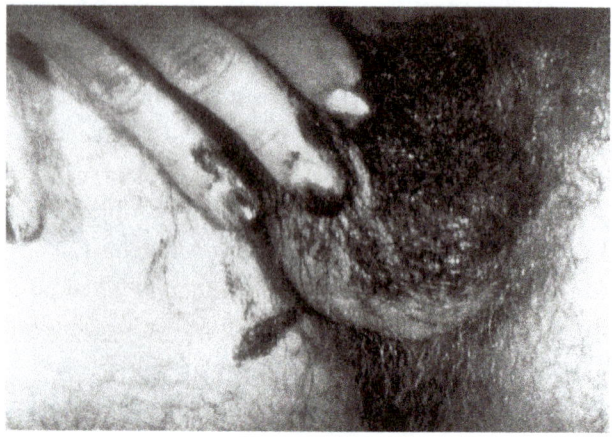

**Figure 10.1** Bruce Nauman, *Black Balls* (1969) © Bruce Nauman/2014, Pro Litteris Zürich.
16 mm film (transferred on video), black and white, silent, 8 min.

The private performance conducted in the confines of his studio was originally captured on 16 mm film and subsequently transferred onto video, which facilitated its presentation in a seemingly endless loop (Osswald 2003: 68). Nauman's decision to use film was probably due to his interest in the extremely slow-motion imagery he wanted to achieve. Video had not yet been technically advanced enough for such a special effect, whereas the industrial camera allowed him to capture up to 4,000 images a minute. Together with *Bouncing Balls* (1969), *Gauze* (1969), and *Pulling Mouth* (1969), *Black Balls* forms a group, all of the films featuring the artist's manipulation of his mouth or testicles in extreme close-up and slow motion (see Schimmel 1994: 73). In *Black Balls* the camera is focused on Nauman's crotch and the finished film shows uncut footage taken from this unchanging vantage point. Nauman's only activity is the handling of his testicles with his right hand. He massages, pulls, deforms, and weighs his private parts, while at the same time dyeing them black with the theater makeup he has on his fingers. It actually took only about twelve seconds to produce the film, but this time-span is extended to eight minutes. Sometimes the action is so decelerated that no motion at all is perceptible—the image just happens to be different from time to time. *Black Balls* has no narrative thread, no suspense, no climax, there is no logical beginning and end to the film.

The artist Bruce Nauman is unrecognizable due to the alienating effects of camera perspective and close-up. The detail of his body and the hand executing the task are separated from his persona. However, Nauman's body parts on display actually *gain* in material reality through this very focus. One becomes aware of the physical matter presented: Minute characteristics of flesh, its weight being held by the hand, the skin stretching over its form, sinews, veins, pores, even hair follicles are brought into focus. The way this human material reacts toward being covered by a coat of black paint leads to a further concentration on its physical characteristics. Spots of smooth skin are easily dyed black, while elsewhere wiry pubic hair complicates the activity. Nauman's body in its materiality thus becomes much more graspable because it is presented on film. The camera focuses on a depersonalized detail of his body, which is brought into the focus of attention through its technological facility of extreme slow motion. Both the apparatus of the camera and the physicality of the video, which produces the seemingly endless loop of a black-and-white image without any sound, play a vital part in the perception of this production. Their technological characteristics support the physical impetus of Nauman's activity.

At first, one might understand a film such as this as a straightforward physical experiment. The repetitive simple activity concentrated on bodily matter suggests Nauman's detachment from his own subjectivity, an almost scientific interest in the physical handling and reactions of his body. But the seeming impartiality of this task and the apparently unmediated presentation of human matter do not exclude its obvious sexual indications. Right away, we are aware that we are staring at Nauman's most intimate body parts. Our gaze is coerced into complete focus by the extreme slowness of the film. We want to identify the physical changes of the image and in doing so stare intently at the physicality of Nauman. Our interest in the visual medium and its characteristics is painfully redirected to the physical matter that this medium explicates.[7] This leaves us, as we are quick to acknowledge, in the uncomfortable role of the voyeur (Dobbe 2003). But what are we looking at exactly? The private activity of masturbation and its public displays in the forms of exhibitionism or pornography come to mind— even more so since the video monitor frames Nauman himself and his audience as its "mirror"-images. Here, Anne M. Wagner's (2000: 68) critique of Krauss's description of video as defined by self-absorption and hermeticism proves correct: "The technology of the monitor opens outward, as well as in. Not only does it register a process of surveillance, it itself asks for monitoring." However, Nauman has not only artificially— on the level of the video—slowed down his activity to almost nothing; he also physically—on the level of the performance—never changes his rhythm. What is more, the film is silent, no soundtrack helps to pigeonhole its narrative line into that of sexual promiscuity.

The ambivalence of perception leads to a strong awareness of one's own role as an observer. Watching Nauman's video is an individual experience, contingent on gender, age, and sociocultural background. The artist's activity—oscillating between private onanism and public exhibitionism, individual interest and objective experiment—holds any number of different roles for his audience in store. The viewer is voyeur, scientist, lover, for example, and always fellow human being with similar physical characteristics and sensual/sexual yearnings. We seesaw between excitement, embarrassment, frustration, and boredom.

## Media matter: A body's semantics and a video's materiality

The presentation of a naked human body always carries a number of narrative implications that no viewer can circumvent. How and why does the human body—especially in its most natural form, naked, without any props or costumes—represent so much, carry so much meaning? And in what ways do performance artists of the late 1960s and early 1970s make use of this immanent mediative force of the human physis?

Erika Fischer-Lichte, in her study on the *Aesthetics of the Performative*, discusses the efforts of performance art of the 1960s as a movement away from meaning and toward effect. The performer confines him- or herself to the production of materiality and sensuality, the effect of which leads the individual observer to generate new meanings (Fischer-Lichte 2004: 242). The focus on the matter of the human body—Fischer-Lichte calls it the "phenomenal body" (phänomenaler Leib) in contrast to the "semiotic body" (semiotischer Körper)—does not lead to a loss of semantization, however. According to Fischer-Lichte, the body's materiality converges with both signifier and signified. Nauman's massaging finger, then, is not de-signified by the artist's bodily fragmentation and the extreme deceleration of his action. Instead, the self-referentiality of this gesture becomes palpable.

Deliberately accepting the tautology of her declaration, Fischer-Lichte (2004: 244–5) understands materiality as the signified, and that signified is for the observing subject always immanent within materiality. The human being *is* a body (Leib) and he or she *has* a body (Körper), which can be manipulated, instrumentalized, and employed to represent (Fischer-Lichte 2004: 129).[8] This twofold specification of the body in a performance video such as Nauman's has consequences for the viewers' reactions. Marie-Luise Angerer (2006: 243) argues— and I agree with her—that even before the human body becomes an agent of symbolic content, gender, or history, it is in itself meaningful, since it codifies the very concept and convention of an individual, self-determined, and conscious human life. If we see a human body we see at the same time our own body and with that ourselves; it is the most classic example of Jacques Lacan's idea of the gaze: the subject (the viewer/audience/us) not only looks at the object (performer), it is simultaneously looked at by the object. The subject is arrested by the object, "pictured by its gaze," as Hal Foster (1996: 108–9) puts it: "The gaze [of the object] is given a strange agency here, and the subject is positioned in a paranoid way." The reality of the body as artistic object is thus indistinguishable from its mediality–but not because it serves right away as an interface between corporeality and culture. Even before the layers of cultural content are inscribed onto the naked body, it already mediates between the performer and the viewer by means of mutual recognition. According to James Elkins (1996: 137), "pictures of the body elicit thoughts about the body, and they can also provoke physical reactions *in* my body."

Nauman documents his activity by means of a video and technologically denaturalizes his body image. At the same time the video as a technology allowing instant feedback conjures up the concept of physical immediacy—Nauman could be acting out his performance at the very time when his contemporary audience is watching the video. According to both McLuhan and Krauss, the all-pervasive if invisible background of video during the early 1970s is the proliferation of mass media. If commercial television is conceived as unapologetically one-directional, positing the active sender against a passive receiver, Nauman seems to react to this norm—he shocks the observers out of their complacent consumption by forcing them to visually consume his private parts.[9] The physicality of his performance, enhanced by the physicality of the documenting medium, necessitates a physical reaction from the audience.[10] This reaction does not necessarily tie authenticity to immediacy. We might immediately close our eyes on Nauman's balls, or have an erection, or both—but we can just as well stare at the loop of his narrow activity until fatigue and hunger become overwhelming. The video and its incentive to react will not stop.

## Vito Acconci's shoot

Vito Acconci's now infamous performance *Seedbed* (1972) is conceptually related to this "dispositiv" of video art—the usage of bodily subject matter or the employment of technological features such as instant feedback in order to imply physical closeness and generate bodily reactions on the part of the perceiver (Figures 10.2 and 10.3).

**Figures 10.2 and 10.3** Vito Acconci, *Seedbed* (1972) © 2014, Pro Litteris Zürich.
Photographs of performance/installation, 9 days, 8 hours a day, 3-week exhibition, Sonnabend Gallery, New York.

Video as "dispositiv" may be understood not as a clearly distinguishable aesthetic category, but instead as a form of knowledge production that, despite, or because of, its historical and discursive malleability, remains a solid and insurmountable factor. Helmut Draxler (2010) utilized the Foucaultian concept of the "dispositiv" for his discussion of painting in contemporary theoretical and artistic discourses. In the English version of his text, the term was—in Draxler's opinion misleadingly—translated with "apparatus." The technicality of "apparatus" seems to displace the cultural and social parameters that according to Draxler are responsible for a "dispositiv" to change. However, for the "dispositiv" of video the technical inclination seems fitting, as the apparatus of the video camera acts in parallel with the social apparatus responsible for its utilization in processes of perception and cognition.

From January 5 to 29, 1972 for eight hours every day Acconci responded to the footsteps of visitors at Sonnabend Gallery, New York, by masturbating under a wooden ramp that had been built in as a second floor slanting down toward the entrance. His verbal fantasies and moans of pleasure were transmitted into the space via a sound projection system. Visitors walked on the wooden ramp in an otherwise empty white cube and only gradually became aware of the sound installation and its content. Though the soundtrack gave them the impression that they were witnessing a private sexual activity, they had no idea if the declarations they heard were actually genuine, if they really were the objects of Acconci's desire and whether his masturbation was motivated by their presence. The corporeal presence of the artist remained unclear as it was mediated through the loudspeakers.

*Seedbed* is conceptually related to the live video performance *Claim*, which Acconci realized one year earlier, in 1971 (Figure 10.4). Here, Acconci sat in the basement of the gallery, was blindfolded, and, swinging a metal pole, verbally attacked the audience that was gathered in the white cube above him and looked at the monitor displaying his activity below. Acconci warned the visitors in monotone insistence not to come downstairs. He claimed that he would hit anybody who entered his room as, being blindfolded, he could not discriminate between friend and foe. Acconci's threats to cause physical harm were taken seriously enough for no one to attempt to enter his private basement.[11] The original live video performance lasted three hours. In the course of the broadcast, Acconci talked himself into a frenzy, rhythmically hitting the floor with his pole. At other times the sheer exhaustion of this self-assigned task became palpable. The repetitive mumble of threats turned into a hoarse whisper, his posture slackened, the pole

**Figure 10.4** Vito Acconci, *Claim Excerpts* (1971) © 2014, Pro Litteris Zürich.
Video, black and white, 62:11 min., documentation of performance, 93 Grand Street, New York.

dropped to the floor. The video monitor's grainy, imperfect surveillance acts as a bridge between visitors and artist—Acconci is brought close and at the same time kept apart. Willoughby Sharp ([1974] 1976: 259), who witnessed the original video performance, describes this irritating physical closeness that is due to its very virtuality on the monitor: "As he began the piece and got into it ... he began to strike out with the crowbar and hit the side wall, causing the whole staircase to shake. This was both seen and heard on the TV and felt in the space. He *was* there, somewhere behind the monitor, somewhere through the door next to the monitor."

The performance *Seedbed* is conceptually affected by this "dispositiv" of video. In *Seedbed* Acconci reframes the idea that the mediatized bodily activity of another person, even if it is just a verbally expressed threat, physically affects the individual observer. In *Claim*, the live medium of video separates the physical location of the performance from its broadcast image, while it is simultaneously recorded and displayed. In *Seedbed*, performer and audience share one room, artificially divided by the inserted floorboards. They have moved closer together, their bodies being separated from each other only by the slanted ramp. Acconci's activity, too, has become a different one; he masturbates to the steps above him and moves underneath his audience according to their movements, which are triggered by his voice coming from the loudspeaker in the far left-hand corner of the gallery. Instead of the broadcast video of the artist, the audience now only hears Acconci's voice, which reacts to their images as he conjures them up in his mind. Gloria Moure (2001: 23) claims that Acconci considered the video monitor first and foremost as a "sound box" and only secondly as projector of images. In *Seedbed* he realizes that shift in functionality. The concept of a physical reaction toward another body, however mediatized it is, a concept that during the 1970s adhered to video in its critical distance from TV, serves as the basis of Acconci's performance. *Seedbed* has learned from video that every perception of another human body—be it a direct image, an instantly broadcast video, or merely the transmission of a human voice—is already mediatized, and that such a mediatized body nevertheless has physical effects on each individual that perceives it.

## Borderline: Communication through bodily affinities

Acconci's performance was swiftly and to some extent rightly interpreted as a critique of the institutionalized art scene and its stubborn insistence on the individual male artist-genius. Acconci ironically literalizes the artistic act. All its stereotypes—the perfection of manual work, creative spontaneity, artistic climax, and the discharge of creativity—are filtered through his sexual activity underneath a room full of unsuspecting connoisseurs. What is more, the individual visitor becomes inexorably involved in the articulation of "artistic" meaning. Acconci exaggeratedly performs the interpretive exchange as an erotic intertwining between his desires and those of the visitors. Amelia Jones thus interprets Acconci's performance as a case study for an art that does not function as a "unique object" but as a "distribution system" involving the phenomenological "interchange" of subjects within the social realm. Staging art as a communicative exchange between desiring subjects, Acconci in her opinion not only unveils and undermines the privileged male body of the artist/genius. He also offers a critique of the traditional function of the artistic object by trading it in for "social interchange" (Jones 1998: 105–6).

The body itself then, in its very corporeality, signifies institutional critique. However, Acconci's criticism is more complicated than the switch from artistic object to "social interchange," which Jones praises as a laudable social critique. The exchange Acconci offers is neither equal nor equally informed. He is in control of both his body and the entire situation and he offers only the soundtrack of his activity—which could just as well be recorded. The visitor is left to make sense of this setup, as he or she stares at the white walls and listens to someone else's sexual phantasies. The immediacy of the situation, its actual corporeal "truth" is not evident as long as Acconci does not clarify his position through the loudspeakers. He reflects on his role, task, and place as an artist in society, but at the same time reconstructs his power as creative force with great emphasis. According to him, the visitors "serve[s] as my [his] medium," since they reinforce Acconci's excitement: "The seed planted on the floor is a joint result of my presence and yours." Also, through his activity, Acconci occupies the entire gallery space with his creative output: "Under the ramp: I'm moving from point to point, covering the floor (I was thinking in terms of producing seed throughout the underground area). ... I have to continue all day—cover the floor with sperm, seed the floor" (Acconci 1972: 62). Acconci's institutional critique is thus carried out with the physical authority of the male artist-creator.

Even though the confrontational aspect of Nauman's performance is also not direct, I still do not share Fischer-Lichte's negative assessment of technical and electronic media with regard to their impact on the performative. Fischer-Lichte (2004: 174–5) believes that by dematerializing the human body, these media allow only for an illusion of bodily presence. These "effects of presence" are, according to her, completely detached from an actual, material corporeality. I think that the parallel Nauman and Acconci present between the material and mediative characteristics of both body and video lead to the perception and sensual recognition of an equally strong parallel between the video that shows the artists' body and the body of the viewer.

Anne M. Wagner (2000: 60) states very convincingly that what was performed in performance and what was observed in video are the uncertainties that by 1970 or thereabouts had begun to accumulate around "artist" and "viewer" as art's two essential correlative terms: "In 1970 these categories started to dissolve, with confidence in their old contents slowly leaking away. Artists who used video and performance realized that this leakage could be channeled for use as the subject matter of their art, as well as dramatized as its chief effect"—especially in connection with their own body. The artists work on the physical and psychological borderline between artistic subject and viewing subject, or aesthetic object and objectified consumer. The contradiction between a spectator and his or her mechanical and cultural mediation, which according to McLuhan is numbed by mass media, is provocatively heightened by videos such as Nauman's and Acconci's (Joselit 2000: 51).

Through the presentation of their bodies, Nauman and Acconci provoke a "leakage" between the categories of producer, product, and consumer. For Niklas Luhmann, art is a specific format of communication because of its very material aspects. In his introduction to *Die Kunst der Gesellschaft* (1995: 9), he explicitly states that he is not interested in developing a theory that explains art, but instead in demonstrating and describing the communicative usage of artworks. His interest is the connection between mental systems (systems of consciousness) and social systems (systems of communication), and the function and role art has for these interferences.

According to Luhmann, an individual's perception of the world can only in a very limited way be utilized for social communication, since there exists no reality continuum between the sphere of perception and that of the world (Luhmann 1995: 21). Art, however, is capable of building some kind of connection between these two spheres, simply because it *is* a form of communication (social realm) and at the same time it *draws* from perception (individual/interior realm). However, perception is normally directed toward finding information in an already well-known world. It is challenged by art, which does not follow this pragmatic stance (Luhmann 1995: 27). Art (in the modern sense), according to Luhmann, does not communicate through language and it also is not geared toward containing specific information that needs to be transmitted. In accordance with that argument, Nauman has stated that art generally adds information to a situation, and that it seems reasonable to also make art by removing information from a situation (Tucker [1970] 2002).

Because of these omissions, art according to Luhmann (1995: 42) is an exceptional medium of communication that increases the awareness that communication is at work. In art, the question is thus shifted from the "what" (is communicated) to the "how" (does communication work) by the very act of nonverbal and nonpragmatic communication. It is exceptional in its access to perception because rather than giving information about something, it informs about the processes of communication itself. Still, says Luhmann, it is undeniably communication since—just like language—it is produced, handmade, not natural. Its material capacities can only function as communication since it is not just the natural outer world, but instead artificial, fashioned, designed, and thus ready to be questioned, provoking irritation and reflection.

The question arising from these thoughts is: How does the artwork communicate through its materiality? This question, in my opinion, becomes specifically pressing when art's material is the human body, since the body is generally understood as a referent for nature. Its traditional placement in the realm of the natural and authentic often leads to the misunderstanding that the image of the body is natural as well and as such in opposition to culture. Instead, one should understand the body as the product of a social construction of nature. It is, to use Luhmann's distinction, a medium in and of itself that manifests itself in specific forms—gestures, poses, or habits. The body as medium thus begins an infinite process of meaning-production (Karentzos et al. 2002: 9). Nauman's films clarify that our necessary interpretation of the body is an open field, and that it is defined by the specific cultural, social, sexual, and historical frame of its respective "presence." The individual observer is not arrested in this ambiguous frame of reference, as Kristeva's usage of the term "borderline" suggests, but rather stimulated into constant renegotiations. The "leakage" between the bodies of artist, artwork, and viewer—this specific form and format of communication—has no specified meaning but is reconstructed depending on one's individual physical affinities.

# Notes

1   This point is of great importance for Krauss, who utilizes her analysis
     of video as a critique toward formalist criticism, which is focused on
     the formal characteristics of each artistic medium: pigment-bearing
     surfaces for painting, matter extended in space for sculpture, etc.
     Video, whose medium-specificity she describes as a psychological state,
     can act as a critique of such formalist reduction. Krauss's positioning
     of video against formalism is an indication of her post-Greenbergian
     attitude within the 1970s' art historical discourse. But even if it is a tool
     to break away from formalism, her concentration on the psychological
     model should still not be disregarded. Verena Krieger offers an
     analysis of Krauss's postmodern critique, which proves to be highly
     informed by modernist formalism (Krieger 2008: 143–61). For Krauss's
     application of psychological models as a critique of formalism, see also
     Krauss 1993.
2   "The demand for instant replay in the media—in fact the creation of
     work that literally does not exist outside of that replay, as is true of
     conceptual art and its nether side, body art—finds its obvious correlative
     in an aesthetic mode by which the self is created through the electronic
     device of feedback" (Krauss 1976: 59). Krauss's narrow definition of

video as the "aesthetic of narcissism" has received many critical readings (Bellour 1988: 327–87; Ross 1996; Wagner 2000: 59–80).

3  Ursula Frohne presents a similar argument for photography, see Frohne 2002, pp. 401–26.

4  See, for example, Nochlin 1988 or Pollock 1990. For a survey of feminist art historical scholarship see Söntgen 1996; Angerer 1995. A more recent feminist approach that utilizes concepts of the postcolonial is presented in Karentzos 2002.

5  Hal Foster agrees with me on this issue (Foster 1996). The inappropriate amalgamation of these concepts was also evident in the planned but unrealized exhibition project "From the *Informe* to the Abject," initiated by Claude Gintz for the Musée d'Art Moderne de la Ville de Paris. Rosalind Krauss (1996: 90) critiques the show's concept in her article "*Informe* without Conclusion": "… To state why and in what way it (abjection) must be differentiated in the strongest possible terms from the project of the informe." This is also the argument of the exhibition *Formless*, which Krauss curated together with Yve-Alain Bois four years later at the Centre Pompidou in Paris (Krauss and Bois 1999).

6  As a "cool medium," the body's information is not "high definition"— well defined, sharp, solid, and detailed. Quite to the contrary, it provides little information, challenging the respondent to individually fill in narrative and semantic vacancies. Terrence Gordon in his introduction to the critical edition of McLuhan's *Understanding Media* comments on McLuhan's differentiation between "hot" and "cool" media. He stresses that when McLuhan speaks of a medium transmitting information, he is not referring to facts or knowledge but to the response of our physical senses to the medium. "Cool media," which provide little information, are thus high in participation (Gordon 2003: xvi–xvii).

7  Nauman fittingly compared this film with documentary footage of the explosions of atomic bombs, "where you become fascinated with the weightless cloud formations unfolding like giant flowers." but are at the same time aware of their morbid implications (Morgan 2002: 57). Bruce Connor's film *Crossroads* (1976) comes to mind, for which Connor used US government archival footage of the first nuclear test series of the postwar era at Bikini Atoll on July 25, 1946. Connor takes the viewer through the atomic bomb test twenty-seven times, from various camera perspectives and at different speeds. Through the permanent, rhythmic

repetition, the explosion is robbed bit by bit of its historical reality, but despite its abstracted beauty its horror remains.

8   Fischer-Lichte discusses this tension between phenomenal and semiotic body for the actor in the theater: "The human being has a body, which he or she can manipulate or instrumentalize just like any other object. At the same time he or she is that body, the human being is a body-subject (Leib-Subjekt)." (my translation) (Fischer-Lichte 2004: 129).

9   For commercial television as a norm of media consumption, against which video art reacts, see David Antin, who makes the connection to McLuhan's positive and Krauss's negative assessment of mass media as well (Antin (1975) 1986: 147–66).

10  This demand of interaction needs to be understood in its historicity just as much as the claim that video art fails to achieve an actual interplay between sender and receiver. Today, video pieces such as Nauman's *Black Balls* cannot be mistaken for actual but remote live performances anymore. Their documentary status is much more evident, due to the outdated technology and its aesthetics as well as the age of the artist. For the subsequent historization and institutional "taming" of video see Rosler 1990.

11  The visitors did have the opportunity to enter the room in which Acconci was situated, but according to Yvonne Spielmann not one person actually went down the stairs to meet his threat (Spielmann 2005: 140). Anja Osswald, however, founding her analysis on the video documentation of the live performance of *Claim*, disagrees with this account and asserts that actually three visitors did walk down the stairs, causing even more verbal violence and threat from Acconci (Osswald 2003: 157–8).

# References

Acconci, Vito (1972), "Script: Seed Bed," *Avalanche*, 6: 62.

Angerer, Marie-Luise (2006), "Performance und Performativität," in Hubertus Butin (ed.), *DuMonts Begriffslexikon zur zeitgenössischen Kunst*, 241–54, Köln: DuMont.

Angerer, Marie-Luise, ed. (1995), *The Body of Gender—Körper, Geschlechter, Identitäten*, Vienna: Passagen-Verlag.

Antin, David ([1975]1986), "Video: The Distinctive Features of the Medium (1975)," in John Hanhardt (ed.) *Video Culture. A Critical Investigation*, 147–66, New York: Visual Studies Workshop Press.

Bellour, Raymond (1988), "Autoportraits," *Communications*, 48: 327–87.

Dobbe, Martina (2003), "Video - the aesthetics of voyeurism? Zur medialen Struktur des Blicks in der frühen Videokunst," in Lydia Hartl, Yasmin Hoffmann, Walburga Hülk, Volker Roloff (eds), *Die Ästhetik des Voyeur/ L'Èsthétique du Voyeur*, Heidelberg: Winter.

Draxler, Helmut (2010), "Painting as Apparatus. 12 theses," *Texte zur Kunst*, 20 (77): 107–10.

Elkins, James (1996), *The Object Stares Back. On the Nature of Seeing*, San Diego, New York and London: Harvest Book.

Fischer-Lichte, Erika (2004), *Ästhetik des Performativen*, Frankfurt/Main: Suhrkamp.

Flach, Sabine (1994), *Körperszenarien. Zum Verhältnis von Körper und Bild in Videoinstallationen*, Munich: Fink.

Forte, Jeanie (1988), "Women's Performance Art," *Theatre Journal*, 40 (2): 217–35.

Foster, Hal (1996), "Obscene, Abject, Traumatic," *October*, 78: 106–24.

Frohne, Ursula (2002), "Berührung mit der Wirklichkeit. Körper und Kontingenz als Signaturen des Realen in der Gegenwartskunst," in Martin Schulz, Dietmar Kamper, and Hans Belting (eds), *Quel Corps? Eine Frage der Repräsentation*, 401–26, Munich: Wilhelm Fink Verlag.

Gever, Martha (1990), "The Feminism Factor: Video and its Relation to Feminism," in Doug Hall and Sally Jo Fifer (eds), *Illuminating Video. An Essential Guide to Video Art*, 226–41, New Jersey: Aperture/BAVC.

Gordon, W. Terrence (2003), "Introduction," in Terrence Gordon (ed.), *Marshall McLuhan. Understanding Media. The Extensions of Man* (1964), xi–xxi, Berkeley: Gingko Press.

Jones, Amelia (1998), *Body Art. Performing the Subject*, Minneapolis and London: University of Minnesota Press.

Joselit, David (2000), "The Video Public Sphere," *Art Journal*, 59 (2): 46–53.

Karentzos, Alexandra Birgit Käufer, Katharina Sykora, Alexandra Karentzos, eds (2002), *Körperproduktionen. Zur Artifizialität der Geschlechter*, Marburg: Jonas Verlag.

Krauss, Rosalind (1976), "Video: The Aesthetics of Narcissism," *October*, 1: 50–64.

Krauss, Rosalind (1993), *The Optical Unconscious*, London and Cambridge, MA: The MIT Press.

Krauss, Rosalind (1996), "*Informe* without Conclusion," *October*, 78: 89–105.

Krauss, Rosalind and Bois, Yve-Alain (1999), Formless: A User's Guide, New York: Zone Books.

Krieger, Verena (2008), "Der Blick der Postmoderne durch die Moderne auf sich selbst. Zur Originalitätskritik von Rosalind Krauss," in Verena Krieger (ed.), *Kunstgeschichte & Gegenwartskunst. Vom Nutzen & Nachteil der Zeitgenossenschaft*, 143–61, Cologne: Böhlau Verlag.

Kristeva, Julia ([1969] 1980), *Desire in Language. A Semiotic Approach to Literature and Art*, 4th edn, New York: Columbia University Press.

Kristeva, Julia (1982), *The Powers of Horror*, trans. Leon S. Roudiez, New York: Columbia University Press.

Lechte, John (2012), "Art, Love, and Melancholia in the work of Julia Kristeva," in John Fletcher and Andrew Benjamin (eds), *Abjection, Melancholia and Love. The Work of Julia Kristeva*, 24–41, London and New York: Routledge.

Luhmann, Niklas (1995), *Die Kunst der Gesellschaft*, Frankfurt/Main: Suhrkamp.

McLuhan, Marshall ([1964] 2003), *Understanding Media. The Extensions of Man*, ed. W. Terrence Gordon, Berkeley: Gingko Press.

Morgan, Robert C., ed. (2002), *Bruce Nauman*, Baltimore: Johns Hopkins University.

Moure, Gloria (2001), "From Words to Things," in Gloria Moure (ed.), *Vito Acconci*, 9–57, Barcelona: Ediciones Polígrafa.

Nochlin, Linda (1988), *Women, Art and Power and Other Essays*, New York: Harper and Row.

Osswald, Anja (2003), *Sexy Lies in Videotapes. Künstlerische Selbstinszenierung in Video um 1970. Bruce Nauman, Vito Acconci, Joan Jonas*, Berlin: Gebrüder Mann Verlag.

Peltomäki, Kirsi (2007), "Affect and Spectatorial Agency: Viewing Institutional Critique in the 1970s," *Art Journal*, 66 (4): 36–51.

Pollock, Griselda (1990), "Missing Women—Rethinking Early Thoughts on Images of Women," Carol Squiers (ed.), *The Critical Image. Essays on Contemporary Photography*, 202–19, Seattle: Bay Press.

Rosler, Martha (1990), "Video: Shedding the Utopian Moment," in Doug Hall and Sally Jo Fifer (eds), *Illuminating Video. An Essential Guide to Video Art*, 31–50, New Jersey: Aperture/BAVC.

Ross, Christine (1996), *Images de surface: L'art vidéo reconsidéré*, Montréal: Artextes.

Schimmel, Paul (1994), "Pay Attention," in Joan Simon (ed.), *Bruce Nauman*, 69–82, Minneapolis: Walker Art Center.

Sharp, Willoughby ([1974]1976), "Videoperformance," in Ira Schneider and Beryl Korot (eds), *Video Art. An Anthology*, 252–67, New York and London: Harcourt Brace Jovanovich.

Söntgen, Beate (1996), *Rahmenwechsel. Kunstgeschichte als feministische Kulturwissenschaft*, Berlin: Akademieverlag.

Spielmann, Yvonne (2005), *Video. Das reflexive Medium*, Frankfurt/Main: Suhrkamp.

Tucker, Marcia ([1970] 2002), "PheNAUMANology (1970)," in Robert C. Morgan (ed.), *Bruce Nauman*, 21–7, Baltimore and London: The Johns Hopkins University Press.

Wagner, Anne M. (2000), "Performance, Video, and the Rhetoric of Presence," *October*, 91: 59–80.

# 11

# The Romantic Readymade: Toward a Material Vitalism of Contemporary Art

Stephen Zepke

The relationship between matter and thought, nature and reason, or body and mind is a perennial philosophical conundrum that continues to animate contemporary debates, not least those around art. The shared conundrum is whether and how the opposed terms can remain distinct, or not. Enlightenment Rationalism forcefully presented and resolved this problem by claiming thought was autonomous and exercised according to its own universal rules and assumptions that explained everything in nature. Simultaneously, romantic thinkers were inspired by vitalism and Spinoza to argue that the universal rules governing thought were the same as those determining relations between all bodies, and so were not those of rational thought, but of organic life itself.[1] One of the by-products of this debate was the question of the function of art. In the romantic tradition it had a privileged place because of its expressive power that directly embodied the univocal vital force through its specific sensory affect. In this way art exemplified the sensuousness of life, and offered a model for its creative materiality. Obviously, Rationalism did not deny sensual experience, but it did claim that this experience was predetermined by a conceptual framework that made it possible. Our experience of causality, for example, was determined by the logical necessity (rather than empirical habit) of cause and effect. Today, the debates between rationalist and vitalist understandings of life seem to continue unabated. In a strange reversal of eighteenth-century events, the recent theory of accelerationism has argued for "the completion of the Enlightenment project of self-criticism and self-mastery, rather than its elimination."[2] While its claims for "futurity" gain substance in its radical claims for a cybernetic consciousness and its "Promethean" desire for a noncorporeal, global, and (in)human reason, its philosophical heritage (emerging from Badiou and Speculative Realism rather than its references to Deleuze and Guattari, Lyotard, and Camatte)[3] suggests a replay of old *ressentiments*, and an inevitable turn against the recent ascendancy of vitalist-inspired philosophy and politics. Given that one of its own conditions of possibility is capitalism's need for palpably "new" products, accelerationism's synergy with contemporary art (which shares this fashion-logic) is hardly surprising. Both are based on a decentering of corporeal experience in favor of a conceptual and discursive mode of production that positions itself within the technological info-networks that increasingly determine our lives. In doing so, the neo-Rationalism of Accelerationism and contemporary artistic practice do not reject the body and aesthetics, but subordinate them to the authority of cybernetic consciousness.

These are obviously broad and no doubt oversimplified claims, and cannot be thoroughly defended here. Instead, these wider issues will be refracted through a very specific lens, that of Duchamp's readymade, and the *gray matter* it privileges as the creative mechanism of art. In fact, our lens will be even more narrowly focused on Deleuze and Guattari's "reading"—although perhaps *detournement* is more accurate—of the readymade, which offers an updated vitalist aesthetics that challenges the "conceptualization" of creative production by accelerationism and contemporary art.

"Representations," Deleuze and Guattari say, "are bodies too!" (1987: 95) This cheerfully pithy slogan immediately flags their materialism, and suggests their vitalist commitments. Succinctly put, it means that representations are real in the same way everything else is, their *corpo*-real, and they "communicate" by affecting other bodies. Affects, in this sense, are dynamic relations defining the intensive state of assemblages, and of relations between assemblages, from the molecular scale all the way up to the cosmic assemblage itself. Now, while this view refuses to privilege the human organism, or even the organism per se, it does privilege aesthetics as the realm of sensation and the affect. Such a view is clearly problematic within the context of contemporary artistic practices that have, since the late 1960s (at least), moved art toward conceptual and discursive modes of production that supposedly gain a critical leverage on the expanding realms of capitalist biopower, that of the "info-commodity," "mass media", and "immaterial labor." In this sense, contemporary artistic practices are "postconceptual" (i.e., their organizing element is conceptual), and explore a "post-aesthetic poetics" (Osbourne 2013: 10 and 33).[4] These inheritances from Marcel Duchamp's readymade attack painting via a "nominalist" understanding of art, turning art into a conceptual "decision" (that something can be named "art") liberating it from any medium-specificity, or aesthetic necessity.

Despite the ubiquity of conceptual artistic practices, Deleuze and Guattari maintain the almost unthinkable position (as unthinkable in 1990, when they were writing, as it is today) that conceptual art is not, in fact, art. Why not? Because, they argue, it dematerializes art by turning it into discursive "information," making its status as "art" depend on the "opinion" of the "general public" as to "what is art?" (Deleuze and Guattari 1994: 198). As a result, art stops being defined as an aesthetic experience, to become an act of communication or a conceptual operation depending upon, and sometimes exploring its linguistic conditions of possibility. The art work thereby becomes an empty "signifier" for a dematerialized signified ("art"), confirming the postmodern truism that reality is constituted by its representation, and that all experience is merely the "reading" of a "text."

Against these conditions of contemporary life and art, Deleuze and Guattari will have recourse to some terribly reactionary positions; on the one hand, old-fashioned philosophy will have to defend the purity of the concept against its ravishing by the market (1994: 10), and on the other, a modernist aesthetics of abstraction and immanent critique will produce a strangely romantic formulation of sensation or affect as embodying the indiscernibility of nature and art. As a privileged expression of the forces that shape the world, art is, for Deleuze and Guattari, an important (perhaps the most important) force of intervention, engagement, and change, but it is so because of its aesthetic and material power, making "political art" an affective rather than a cognitive/critical practice.

In fact, affective political art will finally be, in Deleuze and Guattari's account, a form of animal behavior, making them the perverse inheritors of *Naturphilosophie* and Romanticism inasmuch as art is the materialization of the vital creative process itself. In this sense, the art work embodies what Deleuze and Guattari call "a material vitalism that doubtless exists everywhere" (1987: 411). Material vitalism was, of course, one of the founding claims of the Romantics—who called it *hylozoism*—and for whom matter consists of a self-organizing living force. Nevertheless, Deleuze and Guattari develop a perverse, Nietzschean form of hylozoism in which living force goes beyond any organic teleology, in a process of creative destruction by which life escapes its capture in any fixed form. Nature, in other words, is not purposive, whether in the regulative idea of organic unity discovered by Kant, or in the genetic but nevertheless teleological living force animating organic matter championed by *Naturphilosophie*. For Deleuze and Guattari, nature is a continual and nonteleological process of material vitalism, where the living force animating matter is resolutely inorganic, and its genetic and *a priori* principle is the overcoming of all forms of life. Deleuze and Guattari's transcendental and material vitalism is therefore an ontology of becoming, as it was for Romanticism, but because matter expresses its constitutive and ontological difference, this becoming is embodied in an always already open—rather than organic—whole. This is the meaning of their seemingly romantic claim in *What is Philosophy?* that "Art wants to create the finite that restores the infinite." (1994: 197) Art is the passage from an actual finite to a transcendental infinite, but the infinite is neither an idea nor an organic whole, but a material flux of constant variations or "infinitely varied infinities" (1994: 181). As a result, the finite art work does not simply express an immanent and genetic teleological principle, because this principle is an undetermined virtual event that can only be embodied in a new universe. This then is Deleuze and Guattari's perverse resolution of Kant's and Romanticism's underlying dilemma; art unites the faculties of the imagination and reason (or of the actual and the virtual as they call them) in their sublime and transcendental difference,[5] making vision and contemplation the embodiment of an event that exceeds its conditions of possibility (whether conceptual or organic) by creating a new universe; "Everything is vision," they ecstatically announce, "Everything is vision, becoming. We become universes" (1994: 169). This utterly nonrepresentational aesthetics refuses to bridge the faculties through any *sensus communis*, insisting instead that sensation is nothing less than the eruption of transcendental difference as the real condition of an experience taking us beyond the human, all too human. For Deleuze and Guattari, art therefore retains its romantic naturalism and mystical power,[6] inasmuch as the artist creates nondiscursive sensations expressing this metaphysical (but thoroughly immanent) force, but without any trace of Romanticism's teleological organicism. As we shall see, self-organization is not, for Deleuze and Guattari, a teleological force explaining the organism, but a cosmic power overcoming all limits. Deleuze and Guattari's Romanticism is then, the "cool Romanticism" suggested by Paul Klee, in which the experience of the absolute immanence of nature and art qua "inorganic life" is rinsed of the psych-agonies of *Sehnsucht*.[7] But after all of this Deleuze and Guattari will claim a quite surprising figure as the paradigmatic artist; they will claim as their hero, and not by accident, Marcel Duchamp and his readymade. The ironies of this argument will not escape us, as they are no doubt made for our enjoyment.

In *A Thousand Plateaus* Deleuze and Guattari claim that "Territorial marks are readymades," (1987: 349, see also 1994: 184) using the English word "*readymade*" in order to emphasize its origin in Duchamp. (Deleuze and Guattari 1980: 389, 1991: 174)[8] These readymade "territorial marks" operate in a similar way to the accidental and "catastrophic" marks that open up Francis Bacon's diagram, both turn materials away from their given function and sense to allow them to construct and therefore express something new. As such, both Bacon's painting and Duchamp's readymade operate as what Guattari calls "components of passage" (2011: 138), facilitating the univocal and material movements of de- and re-territorialization that animate both art and life. The similarities and differences between Bacon and Duchamp are important, because they encapsulate on the one side the transcendental aspect of genesis, and on the other very different "styles" of creation. Bacon's approach begins from a violent, unconscious, and physical gesture of abstraction (the random application of paint Bacon calls an "accident" (Deleuze 2003: 94ff)) that wrenches the canvas out of the received world of cliche and opinion that act as its conditions of possible experience (and that the mass media delivers directly into our head) and into a chaotic genetic process. The readymade, on the other hand, and as we shall see in more detail, emerges when an existing object is used in a new way in order to catalyze a *biological* process opening up living and material systems (Deleuze and Guattari will call them territories or assemblages) onto an outside in order to enable them to reproduce and proliferate. The readymade therefore has both a specific function within biological systems (e.g., as part of a reproductive process), and an aesthetic function of expression that is undetermined by its conditions, and whose affects can catalyze potentially radical transformations and make them proliferate.

This contrast of Bacon's painting technique and Duchamp's readymade has interesting implications for us. In one sense it does perpetuate the rather hackneyed opposition of the unconscious and antirational gesture of painting versus the strategic use of tools for conscious ends. We could summarize this opposition in Duchamp's famous words; "Stupid as a painter." On the other hand, however, Bacon's gesture is aimed precisely at the biopolitical mechanisms that constitute our "control society," employing its violent but nevertheless strategic affect to expel the conformity and passivity of consensus to create an aesthetic experience that escapes them. The readymade also begins from this shared world of images and objects, but attempts to singularize them, to turn them into aesthetic "machines" (the term is

Guattari's, and we will have to see what he means by it) that proliferate their singularity. This singularity derives from the aesthetic dimension of these images and objects, while retaining a functional purpose (e.g., reproduction).

The fundamental artistic gesture of the readymade, Deleuze and Guattari argue, is the appropriation of something that is used in a completely different way from its original function. Deleuze and Guattari's example is the stage-maker bird that turns over fallen leaves to mark out the "stage" on which it sings, composing "a complex song made up from its own notes and, at intervals, those of other birds that it imitates." In this the stage-maker bird is a "complete artist" (Deleuze and Guattari 1994: 184), inasmuch as the readymade is already, and from the beginning, "the base or ground of art. Take anything," Deleuze and Guattari say, "and make it a matter of expression" (1987: 349). In this sense, the readymade is a material object used to create an affect, a "matter of expression," or "particle-sign" as Deleuze and Guattari variously call it, which opens the stage-maker's territory onto its outside, an act having both functional and aesthetic effects. But if the functional purpose of this "sign" is to perpetuate the species, its aesthetic dimension opens up the "artist's" territory to the infinite and chaotic forces of the cosmos. In one sense this reading of the readymade is familiar, inasmuch as it makes the gesture of appropriation the fundamental creative act. Deleuze and Guattari's reading is entirely consistent, then, with Duchamp's quip that a readymade is simply an object that has "changed direction," and the implication that this makes all of us (even the birds) artists. What is quite different, however, is that it is the aesthetic dimension of this act of appropriation that carries its transformative potential, a potential no longer limited to the epistemological and ontological conditions of art, but applying to the very conditions of our existence.

If we examine some of Duchamp's other comments concerning the readymade however, we get a clearer idea of Deleuze and Guattari's apostate reading of it. In *The Green Box* Duchamp argues that the readymade is a "snapshot" or "sign of accordance" between it and the laws governing its choice (1973: 27–8). For Duchamp, this choice is entirely independent of the readymade object, which merely exists as "information" (1973: 32), indicating that a conceptual decision (a "nomination" as he called it) has taken place—"this is art." This decision therefore expresses the epistemological conditions (and most famously their institutional structure, as in the celebrated case of the *Fountain*) that determine this nomination to be "true." This made anyone capable of such a decision into an artist, because art's conditions were now conceptual and epistemological rather than based upon an artistic skill, or aesthetic taste. While it is also true that Deleuze and Guattari's readymade can be actualized and achieved by any thing, the logic of this "democratization" is quite different. Duchamp's readymade rests upon the "visual indifference" of its ideal act of genesis, an act open to all inasmuch as all it required was, as Duchamp put it, a "complete anaesthesia," the complete subtraction of the affect from art (1973: 141). For Deleuze and Guattari however, the readymade is a "refrain," "a kind of *asignifying, behavioural* language" (Guattari 2011: 139, italics added) that expresses the two simultaneous operations of any living system, on the one hand its "territorialization" or the emergence and sustenance of its organizational coherence (e.g., as a species), and on the other its "deterritorialization" or opening onto new existential universes. In this sense the readymade refrain is both "an existential 'motif' (or leitmotiv) which installs itself like an 'attractor' within a sensible and significational chaos" (Guattari 1992: 17) and a "contrapuntal" and "polyphonic" expression establishing "interspecies junction points" (Deleuze and Guattari 1994: 185). These two moments are inseparable, so in establishing an existential "territory" the readymade refrain also "deterritorializes" it by forming relations to its outside, as it must do in order to maintain its existence. But the process of deterritorialization is also aleatory, and always carries the potential for an "absolute" deterritorialization that introduces "cosmic forces" into the territory that carry it away, and make it rejoin its "infinite *symphonic plane of composition*" (Deleuze and Guattari 1994: 185). The readymade refrain is therefore "not a teleological conception" (Deleuze and

Guattari 1994: 185) of nature (unlike *Naturphilosophie*), because the material vitalism of "life" proceeds through this always intwined double movement, producing sensations that at once constitute beings and transform them without determined purpose. Deleuze and Guattari's readymade refrain is nevertheless romantic in a certain sense, because it also understands art as a privileged moment of the living whole that it both expresses, and in so doing constructs. "If nature is like art," Deleuze and Guattari tell us, "this is always because it combines these two living elements in every way: House and Universe, *Heimlich* and *Unheimlich*, territory and deterritorialization" (1994: 186).

Art therefore becomes life in the readymade, which seems to fulfill the ambitions of contemporary art, but at the same time Deleuze and Guattari's readymade denies its conceptual genesis (Duchamp's idea) in favor of a materialist—and modernist—aesthetics of sensation. This is the rather odd relationship of Deleuze and Guattari with contemporary art; on the one hand they entirely affirm both the immanence of art and life, and the political potential of this, but on the other they ontologize politics, basing it on the immanent, aesthetic impulse animating all material force. Deleuze and Guattari, we might say, give contemporary art everything it wants, but do so through everything contemporary art seems to reject.

All of this is more clearly explained by Guattari in his book *Cartographies Schizoanalytique*, where he directly discusses Duchamp's first readymade, the *Bottle Rack*. The readymade technique produces, he claims, a "fractal virtualisation" (2013: 206) of the material object, first of all (and like the stage-maker bird, and indeed Bacon's diagram) by isolating it from its context, and so from any pre-given and self-evident function or significance. Guattari is here drawing on the work of Mikhail Bakhtin, who he always mentions when he talks about Duchamp. This process of isolation produces what Bakhtin calls an "active indetermination" (Bakhtin 1990: 275), one that allows the readymade to adopt eccentric trajectories leading to what Guattari will call the art work's "completion as disjunction" (1996: 166). It is precisely this rupture with its received meaning—what Deleuze calls in relation to Bacon, the "catastrophe"—that turns material expressive, allowing it to undergo a potentially infinite number of incorporeal transformations through which it might become something else. This is precisely how Deleuze and Guattari explain the readymade qua refrain in *A Thousand Plateaus*, where the leaf is physically manipulated, but only in order to turn it expressive, to transform it from a leaf into a stage, from photosynthesis to seduction, from plant to sign. Influenced by Bergson's *Matter and Memory*, Bakhtin understands the aesthetic object qua sign to be both a material body and an image, the first generated by the physical act, and the second by the expressive act of appropriation (or as Guattari has it, "consumption" (1992: 14)), an aleatory moment of intervention initiating the objects' eccentric becoming, and in so doing expressing and (re)constructing the material connections constituting its world. The "artistic act," according to Bakhtin, therefore emerges as the inseparability of a material object and the "meaning" it adopts, a dynamic emergence encompassing object and subject in a singular "existential refrain" (Guattari 1992: 15). This refrain may simply lead to a sexual mate, or another room of the "house," but these deterritorializing movements also potentially lead to the cosmic horizon of a "becoming Universe."

The Bakhtinian understanding of the "aesthetic object" is therefore consistent with Deleuze's insistence that the two Kantian senses of the aesthetic—as a theory of perception and as a theory of art—must come together. As such, the "act" of the readymade is a ubiquitous (and indeed transcendental) function producing affects both in nature and as art. In nature this function can operate to maintain the species (the "stage" and song of the stage-maker bird facilitates reproduction), while as art (and "art begins with the animal" (Deleuze and Guattari 1994: 183)) it operates for itself in an "absolute" way. The inseparability of the two moments means "we no longer know what is art and what nature," because both are "natural technique" (Deleuze and Guattari 1994: 185). The readymade deterritorializes an object so that it may pass to its outside, and while this movement may remain relative to the territory it opens, it can also carry the territory away in a movement where deterritorialization becomes reterritorialized on itself, and "variability" becomes the only norm (Deleuze and Guattari 1987: 101). Returning to the artistic act that Guattari draws from Bakhtin, its creation of a readymade opens up the object to a "*polyphonic*" (1996: 193) or "multiplicating" (2013: 211) process of virtualization, while simultaneously actualizing this process in a "subjectivation." Subjectivation is the deterritorialization of the subject, as it is always in excess of its material, discursive, and cognitive conditions of possibility. The readymade's "mutating becomings" (Guattari 2013: 205) therefore have a political effect, inasmuch as they enable, Guattari claims, the "reappropriation of the means of production of subjectivity" (1996: 198), through the creative act of art. In this sense, the "poetic function" of the readymade is inherently political because it re-singularizes the production of subjectivity through an "aesthetic rupture of discursivity" (Guattari 1987: 86) that frees words or images from their representational function by placing them into a sudden and intimate (i.e., affectual) proximity with the spectator. This intimate affect then takes on "a dimension of autonomy that is of an aesthetic order" as it follows its singular path of individuation, introducing new, aesthetic "coefficients of freedom" (Guattari 1996: 198).

This affect created by the readymade is what Guattari calls a "partial object" (1996: 198) or "enunciative substance" (1995: 26), "an Assemblage of enunciation with multiple heads" (2013: 205–6) composed of material detached from any received "content," which is now constituted by the fluid and in principle infinite affective multiplicities of the spectator. The aesthetic object is in this sense "alive" according to Bakhtin, because its "content" is in a state of becoming, making it a "self-sufficient ... segment of the unitary open event of being" (1990: 275, 306–7).

The political ramifications of Guattari's connection of Bakhtin's aesthetic object to Duchamp's readymade emerge through Deleuze and Guattari's understanding of Loius Hjelmslev's materialist linguistics. In Hjelmslev's terms, the readymade refrain does not dialectically oppose a discursive understanding of art with a materialist one, but rather understands discursive production as an aspect of "material vitalism": "Representations are bodies too!" Hjelmslev argued that a linguistic substance is formed when two deterritorialized material flows come together so that the first defines a content and the second, an expression. This happens in the readymade when the "de-framed" aesthetic object produces aleatory affects in its encounter with the "spectator." These flows of content and expression reciprocally determine each other in a process constituting "figures," or "non-signs," or what Deleuze and Guattari describe as "flows-breaks or schizzes that form images through their coming together in a whole, but that do not maintain any identity when they pass from one whole to another" (1983: 240–1). The readymade in this sense does not reproduce its (discursive) conditions of production, but turns these into a transformative event. This is where it becomes possible for art to achieve the "reappropriation of the means of production of subjectivity" as Guattari so ambitiously claimed, not through claiming the "vertical" authority of cognitive labor (the scientists forging the new cybernetic consciousness) as accelerationism claims, nor through the tired revisionism of critical theory as Peter Osbourne would have it, but through the material vitalism of the readymade and the new subjectivations it produces.[9] This is a "machinic," cybernetic process according to Guattari,[10] but it is one that rests upon the affect rather than the concept as its operative mechanism. In this sense, instead of pursuing a rational cybernetic acceleration of capitalism, Guattari explores what he calls capitalism's "machinic unconscious," a radically inhuman "plane of consistency" on which "abstract machines" utilize readymades to "convey singularity points 'extracted' from the cosmos and history" (2011: 12). This at once historical and ahistorical specificity of the readymade's machinic operation stands against the "abstract universals" (Guattari 2011: 12) of both capitalism (i.e., value) and accelerationism. In this sense it repeats one of the defining features of the rupture between Rationalism and *Naturphilosophie*, the former resting on *a priori* universals defining "man," while the vitalists saw the transcendental force of self-organization as being always already historically actualized (and hence, so was reason). Indeed, Guattari's "machinism" offers to our cybernetic present a politics that is more, rather than less relevant because of its aesthetic emphasis. It is precisely its production of affect that means the readymade is necessarily engaged with the digitech defining our contemporary subjectivities.[11]

Deleuze and Guattari argue that Hjelmslev's break with the signifier in favor of nonsigns causes "content and expression to flow according to the flows of desire" and so "constitutes a decoded theory of language about which one can say—an ambiguous tribute—that it is the only linguistics adapted to the nature of both the capitalist and the schizophrenic flows" (1983: 242–3). The difference between capitalist semiotization and the self-organizing (or as Guattari calls it "autopoietic") readymade is therefore internal to the biopolitical sign, being the difference between a cliché or new axiomatic value (such as the "generic artistic modernism" Peter Osbourne claims is inaugurated by the readymade (2013: 81)) and its production of a singular and unquantifiable quality, or sensation.[12] This places the readymade directly within the contemporary political realm, because it is the vital mechanism (the self-organization of difference) that on the one side capitalism seeks to monetize, and on the other produces expressions that escape and create new worlds. These two processes are, of course, becoming increasingly difficult to distinguish, but it is a mark of Deleuze and Guattari's Utopian Modernism that they insist that art's vital and material powers of abstraction (the readymade qua genetic abstract machine) and sensation remain embedded in but autonomous from capital's apparatus of capture.

In this sense, "Only expression gives us the method" (Deleuze and Guattari 1986: 16), because it is only through expression—as the mechanism by which matter proliferates beyond its conditions—that singularization becomes collective.[13] But this process does not pass from one term to the other, singularization to collectivization, but rather expression escapes semio-capital's control precisely because its singular quality does not stay the same. This is what Guattari calls the "semiotic polycentrism" (1996: 153) or proliferation of the aesthetic sign, what Bakhtin calls the "immanent overcoming" of the material object in its transformation into an aesthetic object or work of art (1990: 297).

The readymade singularizes its object by making it express an affect, but this "act" is infinitely repeatable, a process encompassing the collective in the widest sense. This collective aspect of the readymade is emphasized in Guattari's use of the term "autopoiesis" to describe it, which he takes from the biologists Varela and Maturana's explanation of the entwined development of an autonomous organism and its environment. Autopoiesis emerges as a "reciprocal relation" between local components and their global whole: "An entity self-separates from its background," Varela explains, and in so doing "the autopoietic unity *creates a perspective* from which the exterior is one" (1992: 7). This relation of reciprocal determination, which is "enlarged" by Guattari beyond the limits of the organism to encompass "aesthetic creation" (1992: 93), means that the readymade is not only an expression of its "environment" but also its simultaneous (re)construction through, as Varela calls it, a process of "world-making" (1992: 8). In this it also echoes the aesthetic theories of Romanticism, as Guattari explicitly acknowledges: "The operators of this crystallization are fragments of asignifying chains of the type that Schlegel likens to works of art ('A fragment like a miniature work of art must be totally detached from the surrounding world and closed on itself like a hedgehog')." (2000: 55).[14] These fragments act, as we have seen, as autonomous aesthetic catalysts of existential bifurcations and occupy, according to Guattari, "a privileged position within the collective Assemblages of enunciations of our era" (1992: 101).

Let us go back to Guattari's example of the first readymade, the *Bottle Rack* (1913), to try and see how this works in reality. The *Bottle Rack*, he writes, "functions as the trigger for a Constellation of universes of reference that sets off intimate reminiscences—the cellar of the house, a certain winter, beams of light on the spider webs, adolescent solitude—as much as it does connotations of a cultural and economic order—the epoch in which bottles were still washed using a bottle brush ..." (2013: 209, translation modified). The *Bottle Rack* appears here in the by-now-familiar double register of the "refrain," first of all congealing a personal and immediate assemblage of sensory affects (winter, a ray of light, solitude) that gives a singular and intimate "feeling of being." This ontological singularity then catalyzes an infinity of involuntary memories and more elaborate cognitive processes, inducing, Guattari claims, "innumerable sentimental, mythical, historical and social references" (2013: 205). In this way the refrain produces a "heterogenesis" in which the generation of an infinity of virtual "universes of reference" appears at both an infinite speed, and as infinitesimal deviations in space. The affects generated by the readymade therefore go beyond Kant's transcendental conditions of possibility, and launch us into a transcendental "fractalization" of space and time in and by (real) experience. The readymade, Deleuze and Guattari argue, is a machine that synthesizes material and force, where matter is understood as molecular and force as being cosmic. This defines, they say, our modern age, where such machines have taken the place of the *a priori* categories of space and time as "the ground in a priori synthetic judgment" (1987: 378–9). Similarly, the infinity of heterogeneous durations produced by the readymade go beyond the form-content distinction underlying discursive representational schema, because form no longer expresses content but is constructed by it. "Content" is thereby understood not as something separate from matter, but produced by its dynamic flux and vital force. On the one hand the readymade is a process of "existential grasping" that appropriates material in such a way as to make it open and expressive, giving it a virtual complexity of chaotic proportions, while on the other it develops this complexity within a subjectivation that "decelerates" (Guattari 1992: 114) and actualizes virtual complexity in an aesthetic sensation. The virtual and the actual are in reciprocal presupposition at this point, producing sensations that both express and construct the transcendental realm of becoming (i.e., material vitalism) in and as an actual lived reality. In this sense then, and as Guattari puts it, the readymade "lives under the double regime of a discursive slowing down and of an absolute speed of non-separability" (1992: 115), or in other words (once more Guattari's) it embodies the ethico-aesthetic paradigm of chaosmosis.

Guattari will elaborate this process by evoking "the 'spectator' in Marcel Duchamp's sense" (1992: 14). Guattari's reading of Duchamp's 1957 essay "The Creative Act" from which his idea of the spectator comes is interesting in many respects. On the one hand, Guattari seems to intuitively grasp how this text has a quite apostate relation to Duchamp's other published statements, particularly its quotation of T. S. Eliot's essay "Tradition and the Individual Talent." Indeed, Rosalind Krauss angrily dismissed this quotation of Eliot as Duchamp's "betrayal of Duchamp" (1992: 35). Perhaps for this reason Duchamp's claim that it is the "spectator" who completes the artwork, and contributes what he calls the "art coefficient" that creates "art in its raw state" (1973: 139) does, on the face of it, chime with Guattari's Bakhtinian reading of the readymade. On the other hand, however, Duchamp's quotation of Eliot seems less interested in the artist's relation to tradition than in the separation of an intuitive artistic "feeling" from a creative mental act, the latter belonging to the spectators who finally decide on the art works value because they, and not the artist, has, Duchamp argues, "a state of consciousness on the esthetic plane" (1973: 138). It is precisely this distinction between an artistic "feeling" and the "conscious act" of the spectator that is at stake in the readymade. Contemporary practice seeks to jettison the former while extending the latter to both artist and spectator inasmuch as they are subsumed by the democratized category of "producer." For Guattari, the readymade also empowers the "producer," but in exactly the opposite way, by democratizing the artistic power of "feeling" (or affect). As a result, Guattari's repeated reference to Duchamp's spectator can only be understood as a particularly cheeky example of Duchamp's "Creative Act." Guattari treats Duchamp as a readymade, but as a result it is no longer Duchamp's readymade, which is—and this is the very ontology of its aesthetic being—re-made.

As Guattari says of Duchamp, at once paying him the highest compliment but only by turning him into someone else; "Marcel Duchamp declared: 'art is a road which leads towards regions which are not governed by time and space'" (1992: 101). If we look at what Duchamp actually said we can see the creative perversity of Guattari's quotation, which is perhaps better called a creative appropriation. Duchamp said: "I believe that art is the only kind of activity in which man, as man, shows himself to be a true individual capable of going beyond the animal phase. Art is an opening toward regions which are not ruled by space and time" (1973: 137). Rather than indicating the way that animal refrains of art escape the "preformed structures and coordinates" (1992: 101) of time and space, as Guattari claims, Duchamp seems to be suggesting (in line with the now hegemonic understanding of contemporary art as "postconceptual") that the art object is "an intellectual expression" (1973: 126) of a mental process that seeks to detach itself from the natural world and aesthetics.

Guattari re-makes the readymade so that it becomes a contemporary version of Romanticism's hylozoism, an updated vision of the unity of nature and art. True to its romantic roots this unity provides the ontological ground of aesthetic expression and construction—understood here as the heterogenesis of sensation—and the political mechanism by which singularization can become collective. This should not, once more, be understood as a process that passes from one term to the other, as something singular becoming collective, but rather it is the *production* of singular qualities that takes on collective ownership. This collective means of production is the power to singularize itself, because the readymade expresses and constructs a future that is the materialization of the necessarily collective nature of ontological difference.

Finally then, where does this form of expressive materialism leave us in relation to Deleuze and Guattari's understanding of contemporary art? Certainly, their insistence on sensation as the realm of art does not condemn it to being a historical relic from the age of Kant. On the contrary, understanding art as sensation gives it a direct role in the production of subjectivity, and so places it in a direct relation to the biopolitical mechanisms of contemporary capitalism. But this relation can only be truly transformational, Deleuze and Guattari argue, through the aesthetic autonomy and specificity of art, which insures its incompatibility with the rational and disappointingly human processes of representation and signification. This is where their aesthetics seems to lose touch with much contemporary artistic practice, not to mention current theories such as accelerationism. Art, Deleuze and Guattari insist, is not a matter of information or its communication, it is instead a "vacuole of non-communication," the production of a "non-sense" capable of constructing an outside to our everyday affections and perceptions, to our boring opinions and banal thoughts. Expression is the construction of this outside, of a new future, giving art the utopian function of democratizing the processes of production, making aesthetic creation not only an ontological process but also the basis of any liberatory politics. This is a strangely modernist position, and raises significant challenges for those of us who wish to use it to understand contemporary art. On the one hand, its figures of rhizomes and nomads, and evocations of constant and creative movement seem to appeal to our contemporary experience. But on the other, Deleuze and Guattari's equation of art and nature, and their demand for an a-signifying and noncommunicative expressionist materialism seems hopelessly romantic. But perhaps this is as it should be and as Deleuze and Guattari would have wanted it: a concept of expression that offers us the method of a minor art, one that would not be contemporary at all, but would constitute, as Guattari called it, an "*A-temporal* art, where the cursor of time is brought to the point of the autopoietic nucleus" (2011: 53). The Romanticism of this autopoietic readymade, its hylozoic ontology and aesthetic insistence, seems to remove it from our contemporary context. But it is precisely this process of removal that leads, through the readymade, to a radical revitalization of our existential coordinates, and should not be ignored for more fashionable, but also more compromised, conceptual alternatives.

# Notes

1   For detailed accounts of this essential philosophical conflict see Beiser 1987 and Zammito 1992. For an excellent exploration of romantic vitalism within a Deleuzean frame see Toscano 2006.

2 Williams and Srnicek (2014: 362). Frederick Beiser claims that the single question animating philosophy at the end of the eighteenth century was the "authority of reason" (1987: 1).

3 The *accelerationist reader* includes texts from all three as historical "forerunners." For a vitalist reading of Jacques Camatte, see Zepke 2014.

4 Osbourne does not deny that aesthetic experience remains an aspect of "post-conceptual" art, only that they are subordinated to the conceptual "Idea" conditioning the work.

5 For a more detailed examination of Deleuze and Guattari's sublime conditions of contemporary art, see Zepke 2011.

6 For an extended discussion of Deleuze and Guattari's "mystical atheism" see Zepke 2005, especially chapter 2.

7 Klee's rejection of what he calls the "crass emotional phase of Romanticism," in favor of a "cool Romanticism" or "new Romanticism" rejects the heroic solitude of the romantic artist in order to "embrace the life force itself," at "the source of creation" (1948: 49). Such an art would be "a Romanticism which is one with the universe" (1948: 43) and would find its definitively modern statement in the words Klee placed on his own tombstone: "I cannot be grasped in immanence." Finally, and importantly for Deleuze and Guattari, Klee will advocate a politically engaged and "modern Romanticism" that calls for the creation of a people to come (1948: 55).

8 "Readymade" typically remains in English in French texts on Duchamp.

9 Antonio Negri rightly points out that while for accelerationism "the key issue is to liberate the latent productive forces, as revolutionary materialism has always done" (2014: 368), their "optimistic perception of the technosocial body" (i.e., the supposed separability of cybernetic research and its capitalist exploitation/inhibition) leads to an "underestimation of the social, political and cooperative elements" (2014: 370). Maurizio Lazzarato offers an interesting analysis of the readymade in these terms, arguing that it offers a "third way" by "hybridizing the artist and the technician" (2010: 102), and so escapes "the distribution of competencies in contemporary capitalism" (2010: 103). But Lazzarato's account emphasizes the logical indeterminability of the readymade rather than its affect ("neutralizing" rather than multiplying the object's use and meaning (2010: 110)), in order to offer "a new use of artistic techniques beyond the realm of art" (2010: 101). Lazzarato therefore repeats the tired orthodoxy that art can only be

political (or rather *is* political) when it is no longer art. While Deleuze and Guattari also champion the indiscernibility of art and life, this is because life is *already* art, which in fact provides the model for it.

10   An enlightening account of Guattari's engagement with cybernetic theory can be found in Watson (2009).

11   In this sense, recent analyses such as Joselit (2013) that reject aesthetic singularity in favor of the "buzz" of swarm production (2013: 16) miss a crucial point; on the level of the cybernetic "machinic unconscious" it is precisely the production of aesthetic singularity that embodies a revolutionary *social* becoming, rather than a "new" social "coherence" that remains defined by the market (Joselit 2013: 18). Joselit, like accelerationism, is too enamored of the "anti-art" aspects of distributed digital image-production to appreciate the political potential of autonomous aesthetic affects.

12   Eric Alliez (2013) has recently offered an extremely detailed and subtle analysis of Duchamp's wordplay that locates the readymade as the prophetic locus of and resistance to the flows of semio-commodities constituting contemporary life.

13   No doubt in the forty years since Deleuze and Guattari wrote these words, "the general incitement to expression" (Lazzarato 2010: 101), and its cybernetic collectivization of singularity (i.e., social networking) has become a significant factor in contemporary capitalism's techniques of subjection. While this certainly needs critical analysis, from the perspective of Deleuze and Guattari it amplifies the political possibilities of the readymade, rather than reduces them.

14   Interestingly, Osbourne compares Schlegel on the fragment with Sol Le Witt's "Sentences on Conceptual Art" as part of his wider argument that postconceptual art finds its roots in a post-Hegelian Romantic philosophy of art (2013: 10 and 53ff). A full discussion of this claim will have to wait for another occasion.

# References

Alliez, E. (2013), *Défaire l'image, De l'art contemporain*, Paris: Les presses du reél.

Bakhtin, M. (1990), "The Problem of Content, Material, and Form in Verbal Art," in M. Holquist and V. Liapunov (eds), *Art and Answerability, Early*

*Philosophical Essays by M. M. Bakhtin*, trans. K. Brostrom, Austin: University of Texas Press.

Beiser, F. (1987), *The Fate of Reason, German Philosophy from Kant to Fichte*, Cambridge, MA: Harvard University Press.

Deleuze, G. (2003), *Francis Bacon, the logic of sensation*, trans. D. Smith, London and New York: Continuum.

Deleuze, G. and F. Guattari (1980), *Mille Plateaux, Capitalisme et Schizophrénie*, Paris: Minuit.

Deleuze, G. and F. Guattari (1983), *Anti-Oedipus, Capitalism and Schizophrenia*, trans. R. Hurley, M. Seem and H. R. Lane, Minneapolis: University of Minnesota Press.

Deleuze, G. and F. Guattari (1986), *Kafka, Toward a Minor Literature*, trans. D. Polan, Minneapolis: University of Minnesota Press.

Deleuze, G. and F. Guattari (1987), *A Thousand Plateaus, Capitalism and Schizophrenia*, trans. B. Massumi, London and New York: Continuum.

Deleuze, G. and F. Guattari (1991), *Qu'est-ce que la philosophie?* Paris: Minuit.

Deleuze, G. and F. Guattari (1994), *What Is Philosophy?* trans. H. Tomlinson and G. Burchell, New York: Columbia University Press.

Guattari, F. (1987), "Cracks in the Street," *Fash Art* 135 (summer): 82–5, trans. A. Gibault and J. Johnson.

Guattari, F. (1992), *Chaosmosis, an ethico-aesthetic paradigm*, trans. P. Bains and J. Pefanis, Sydney: Power Publications.

Guattari, F. (1996), *The Guattari Reader*, ed. G. Genosko, Oxford: Blackwell.

Guattari, F. (2000), *The Three Ecologies*, trans. I. Pindar and P. Sutton, London: The Athlone Press.

Guattari, F. (2011), "On Contemporary Art, Interview with Oliver Zahm, April 1992," in E. Alliez and A. Goffrey (eds), *The Guattari Effect*, trans. S. Zepke, London and New York: Continuum.

Guattari, F. (2013), *Schizoanalytic Cartographies*, trans. A. Goffrey, London and New York: Bloomsbury.

Joselit, D. (2013), *After Art*, Princeton: Princeton University Press.

Klee, P. (1948), *On Modern Art*, trans. P. Findley, London: Faber and Faber.

Krauss, R. (1992), *The Definitively Unfinished Marcel Duchamp*, Cambridge, MA: The MIT Press.

Lazzarato, M. (2010), "The Practice and Anti-Dialectical Thought of an 'Anartist,'" in S. O'Sullivan and S. Zepke (eds), *Deleuze and Contemporary Art*, 100–15, Edinburgh: Edinburgh University Press.

Negri, T. (2014), "Some Reflections on the #Accelerate Manifesto," in R. Mackay and A. Avanessian (eds), *#Accelerate, the Accelerationist Reader*, Falmouth: Urbanomic.

Osbourne, P. (2013), *Anywhere or Not at All, Philosophy of Contemporary Art*, London: Verso.

Toscano, A. (2006), *The Theatre of Production, Philosophy and Individuation between Kant and Deleuze*, London: Palgrave.

Varela, F. (1992), "Autopoiesis and a Biology of Intentionality," Available online: ftp://ftp.eeng.dcu.ie/pub/alife/bmcm9401/varela.pdf.

Watson, J. (2009), *Guattari's Diagrammatic Thought, Writing Between Lacan and Deleuze*, London: Continuum.

Williams, A. and Srnicek, N. (2014), "#Accelerate: Manifesto for an Accelerationist Politics," in R. Mackay and A. Avanessian (eds), *#Accelerate, the Accelerationist Reader*, Falmouth: Urbanomic.

Zammito, J. (1992), *The Genesis of Kant's Critique of Judgment*, Chicago: The University of Chicago Press.

Zepke, S. (2005), *Art as Abstract Machine, Ontology and Aesthetics in Deleuze and Guattari*, New York: Routledge.

Zepke, S. (2011), "The Sublime Conditions of Contemporary Art," *Deleuze Studies*, 5 (1): 73–83.

Zepke, S. (2014), "Schizo-revolutionary Art; Deleuze, Guattari and Communisation Theory," in B. Buchanan and L. Collins (eds), *Deleuze and the Schizoanalysis of Visual Art*, London and New York: Bloomsbury.

Part Five

# [Sound|Matter]

# 12

# Revisiting the Voice in Media and as Medium: New Materialist Propositions

Milla Tiainen

## Introduction

Approached with varying attention to its sensory, auditory, and signifying dimensions, over the past decades the voice has attracted repeated investigation in theories and differently focused analyses of media. The voice in media has been conceptualized across a series of undertakings, from film studies to recent explorations of sound in the digital era. Whether overtly or more implicitly, diverse approaches within and beyond media research have also speculated about the voice itself as a medium or as involved in crucial processes of mediation. The voice has thus been implicated in the very concepts of "media" and "mediation"—terms whose uses span a famously multilayered range. Sarah Kember and Joanna Zylinska (2012: 19–21) have recently reminded their readers of the manifold usages of the term "mediation," not just in media studies but also in a number of coexisting domains, from Marxist theory to psychology and sociology. Eva Horn (2007), in turn, refers to the cacophonous understandings and delineations of "medium" as a constituting and fertile feature of media studies. Emphasizing the infeasibility of any fixed definitions and consensus, she asserts how "doors and mirrors, computers and gramophones, electricity and newspapers, television and telescopes … water and air, information and noise, numbers and calendars, images, writing, and voice—all these highly disparate objects and phenomena fall into media studies' purview" (Horn 2007: 7–8).

Existing approaches to voice, media, and mediation provide both the inspiration and insights in need of reappraisal for this chapter. However, the task I wish to engage does not involve attending to all the various ways in which the voice has fallen into the purview of media studies. Revisiting the voice in media with inclusive references to the different media expressions, technologies, and theoretical lines in relation to which it has hitherto been explored, would be impossible for any one essay. Even less viable would it be to comprehensively reassess the previous associations of voice with the concepts of media and mediation across a multiplicity of (inter)disciplinary settings.

The goals of this text are more modest and specific: I want to engage three themes that can be extracted from investigations into voice and media conjunctures. The main claim of this chapter is that a further thinking of these themes can advance our grasp of how voices figure with/in media and how their operations might contribute to understanding media; also, how recent reconceptualizations of media and mediation may help to reappraise voices' manners of existing and the occasions of reality they participate in. The first theme extractable from discussions of voice/media coalescence that I hope to extend concerns the relationality of voice. The second has to do with voice as a sensory, perceptual event. The third and concluding theme addresses the ontological elasticity of voice that refers to its ever-evolving actualizations and reinvented possibilities in mediatized milieus.

The inflections of each theme I intend to propose are incited by lines of new materialist thinking. Only few surveys of sound and next to none focused on the voice have so far drawn on approaches associable with this term.[1] Notwithstanding these hitherto germinal appropriations in the study of sonic and auditory processes, new materialism is, on the whole, an expansively employed concept. It currently characterizes a host of projects and reoriented research agendas in a spectrum of humanities and social sciences, from media and cultural studies to social studies of science, feminist scholarship, and art theory. There is no way of reducing the variegated foci and theoretical affiliations of these forays into a set of determinate, uniformly shared traits. As will be elucidated, I nonetheless find it feasible to excerpt from these projects such conceptualizations and tendencies of rethinking that encourage reconsideration of the themes this essay foregrounds.

As regards the relationality of voice, I will evoke the new materialist emphasis on emergence. At stake is the interminable actualization of reality that derives from connectivity across human and nonhuman, material, social, and semiotic elements and their "open series of capacities or potencies" (Coole and Frost 2010: 10). I will summon the notions of emergence and relation (re-)issued by new materialist projects in order to suggest that the voice may not merely elicit interaction between terms—like self/other, human body/media technology—that are supposedly ontologically distinguishable from their relating. It might be better understood as a catalyst and outcome of *constitutive* relations—relations as the moving ontological or ontogenetic condition for the very existence and becoming anew of entities.

Constitutive relationality is also at stake in reconsidering the voice as a sensory, perceptual event along lines characterizable as new materialist. Of particular import here is the conceptual pair of actual and virtual, which strands of new materialist thought have recalled in order to revisit the sensing body and the relations underlying perceptual experience. The question I will pursue is how these reciprocally presupposing terms might retune our conceptions about the voice's appeal to the sensorium. This question arises in the broader context of film and media studies' renewed engagement with bodily perceptual processes and "deeply sensual, synaesthetic" effects of the media (Beugnet 2007: 2).[2]

My third theme regarding the overall ontological elasticity of voice expands on the notions of emergence and relation. Many investigations within and beyond media and film studies have concerned themselves with the radical technological, historical, and sociocultural mutability of vocal expressions, including their connections to bodies, signification, spaces, and, to a degree, nonhuman things and forces. I argue, however, that when addressing the transmutations of voice, increasing attention should be paid to their occurring on a continuum of relations between human and more-than-human. The latter appellation signals activities and "contingencies belonging to any number of categories" as well as the excess of "currently human potential" (Massumi 2013: xxiii). I will make some propositions toward this shifted outlook by calling for contextually attuned uses of posthuman thinking in voice studies. Following Rosi Braidotti (2013: 1–12) among others, the posthuman designates both a heightened tendency of contemporary realities and a related ongoing reorientation of critical and creative thought. As I will elaborate, for present purposes it is justified to regard new materialist and posthuman theorizations as closely associated bodies of response to similar problems.

This chapter does not approach theoretical reformulation as postulation of (new) generalities detachable from situational detail and conditions of articulation. My enquiries of the first two themes in particular gain impetus and shape from one selected example in each case. To paraphrase Thomas Elsaesser and Malte Hagener (2010: 8), the intention is to highlight the generative interpenetration of media theory and "practice": theory builds on media cultural examples, while the dimensions and capacities of examples proliferate in contact with theory.

Lodged in a predominantly cinematic context, the example fueling my propositions about voice and relationality is the award-winning British feature film *The King's Speech* (Hooper 2010). This movie renders a portrayal of the struggles of King George VI with his vocalic production in the face of required acts of public, media-disseminated speaking. It is, in turn, the vocal expressions of Armenian American singer, composer, and voice experimenter Cathy Berberian that co-construct my reconsiderations of the voice as a sensory, perceptual event. I will attend to Berberian's renowned composition *Stripsody* (1966), which bears a peculiar relation to the visual medium of comic strips. My propositions on the posthuman voice in this text mainly extend the two preceding lines of argument. In the conclusions, I will demonstrate how this third theme grows out of the previous ones, helping to assess their implications and push them forward.

The undeniable difference of my examples in terms of medium and related research debates will be acknowledged as much as the limits of this chapter allow. However, it is their intense focusing on the voice and the interlinked ways in which they encourage new materialist reconceptions of voice, media, and mediation that legitimate the co-figuring of the examples in the present text.

The next three sections of my chapter illuminate some of the central ways in which the problems of relation, sensory experience, and ontological contingency have thus far featured in explorations of media concerned with the voice. The considerations of the previous handlings of each theme extend into preliminary expositions on what new materialist lines of approach may bring into their examination. The two subsequent sections then put those lines in further action through their example-rooted discussions on the essay's themes and the concepts of media and mediation.

## Voice and relation: From a mediator in-between to relational events

Along with such other areas as continental philosophy and feminist criticism, media and film theoretical engagements with voice have repeatedly interrogated the voice's powers to establish relations. To give a recent example, Norie Neumark (2010: xix–xx), in her introduction to the edited volume *Voice: Vocal Aesthetics in Digital Arts and Media*, writes about "alterity." This figure conveys for her one of the key questions that the voice raises not only for media scholars but also across an interdisciplinary span. In Neumark's usage, alterity signals the voice's oft-noted fundamental disturbance of any notions of self-containment. As physical vibration and audible sounds and signs, voices irrevocably exceed the subject or body—whether organic or technological—from which they emanate. While doing so, they by the same token interrelate bodies, subjects, and spatial, socio-material milieus. Paraphrasing Walter J. Ong's early work in sound studies, Neumark thus highlights the voice as a dynamic binding vector "within a total sensual/spatial/temporal situation." She furthermore defines its alterity as performative intersubjectivity: as the sounding out of "the physical, affective, signifying, and psychic spaces between subjects" (Neumark 2010: xix, xvi).

With respect to thematizations of relation in voice-centered investigations of media, Neumark's formulations give rise to two remarks. First, whereas the aforementioned edited volume examines the voice's alterity in various practices of digital media culture ranging from podcasting to Internet voice art, resonating acknowledgments of the dispersive and connective workings of voice can surely be found in earlier analyses of the voice in media. They are discernible in Kaja Silverman's inaugural feminist psychoanalytical study of cinematic voices, *The Acoustic Mirror*. In Silverman's characterization, voice is "capable of being internalised at the same time as it is externalised, to spill over from subject to object and object to subject, violating the bodily limits upon which classic subjectivity depends" (Silverman 1988: 80). Both Silverman's analysis of the gendered hierarchies around voice in Hollywood and experimental repertoires, and Michel Chion's likewise pioneering mapping of cinematic vocalities can, among other things, be seen as charting the specifically filmic renderings of the voice's connective, relational capacities. In Chion's case, this shows, for example, in his psychoanalytically informed discussion of the "nurturing connections" that voices (that are on these occasions often transmitted by media technology) at times provide for filmic characters. It also marks his musings over the acousmatic voice's suffusion of both the film characters' and auditor-spectators' experiential space (Chion 1999: 17–29, 61–66).

The second remark to be made here is that Neumark's associations of voice with interrelatedness and the transgression of customarily assumed borders echo several established notions of media and mediation. There are some resonances to be traced from her notion of alterity all the way back to Marshall McLuhan's (1999: 78–9) postulations about the voice as a medium and, therefore, as an extension of man. Descriptions of the in-between spaces that the voice articulates bear some traces of such traditional understandings of mediation where this term denotes negotiating or intervening "third" factors between entities (Kember and Zylinska 2012: 19).

Now, one of the signal features of new materialist and intellectually proximate projects in the current humanities and social sciences consists of resuscitations and further development of such views of relation that may, alongside their other influence, expand our ideas about the relationality of voices. Through returns to a set of thinkers including Gilles Deleuze and Félix Guattari, William James, Alfred North Whitehead, and Gilbert Simondon, the new materialist approaches initiated over the past two decades typically aim at dispelling an arguably persistent tendency regarding notions of relation that is rooted in the so-called substance metaphysical traditions of Western thought. This is the conception of relation as an occurrence that arrives after and merely supplements the primary individuality of the relating terms.

As a means of reconfiguring this ontological scheme of separateness, we can find in the writings of new materialist theorists—for example, Karen Barad (2007), Rosi Braidotti (2006), Jane Bennett (2010), Manuel DeLanda (2006), and Vicki Kirby (2011)—an array of conceptual suggestions, from Deleuze-Guattari's assemblage to Barad's quantum physics-derived entanglement, that advance an understanding of relations as generative and regenerative of the relating entities' very individuality. To deploy Simondon's vocabulary as another revived mode of conceptualization, relations are thus reconceived as a pre-individual structuring condition for their participants' co-occurring individuations. These are always provisionally attained states of being. Relations are where the entities' qualities, capacities, and self-relations (connections to their respective pasts and unfurling toward the future) jointly emerge and modify.[3]

Despite the varying concerns and theoretical impulses of new materialist projects, their elaborations of the above views involve shared aims. Key among these is to conceptually and analytically reclaim the reality-making role of materialities without assigning them a self-contained identity or ontological primacy. The already-known motto reiterated in many studies that can be called new materialist is that humanistic and social scientific analyses of contemporary reality and renewed considerations of ontology should pay growing heed to forms and modes of materiality stretching from technological systems to natural forces and "evolving corporeal practices" (Alaimo and Hekman 2008: 3). What the constitutive views of relation help to avoid is the positing of these modes as self-standing existential layers or determining factors. Rather, the activities and affectivity of material phenomena arise in complex, mutually inclusive relations with other processes of reality, whether these are social, linguistic, representational, scientific, or artistic.

To return to the questions of this chapter, this concept of relations proposes at least two reconfigured perspectives for the study of voice/ media. In its wake it is not quite adequate to say, as Neumark (2010: xvi) puts it, that the voice sounds out "physical, affective, signifying, and psychic" spaces between subjects and milieus. To an extent, this phrasing still implies preconstituted, only secondarily connected identities. Therefore, the voice would be better regarded as a processual factor that partakes in and initiates relational events where the states and capacities of the involved entities veritably transform—become anew.

Theorizations of relational emergence also enhance understandings of the voice itself as an event, in the sense that as a sonic phenomenon it has no permanence but repeatedly occurs as sounds. This aspect has preoccupied various engagements with voice across areas from philosophy and sound studies to media theory. Following new materialist revivals of relational ontology, vocal expressions necessarily result from the co-constitutive interplay of many elements that are heterogeneous and, hence, also material in kind. These can encompass anything from the materialities and habits of the vocalizing bodies to the media technological and institutional factors of the vocal events. These assumptions have implications for what the voice can mediate and whether or how it may be considered a medium. I will elaborate these questions below in interchange with *The King's Speech*.

## The virtual in the sensory?

Like relation, questions of sensorial experience and perception have figured frequently enough in previous media theoretical explorations of voice. Their importance was first laid down at the formative stages of these enquiries. This is evident in Chion's *The Voice in Cinema* when he announces the focus and ambition of his examinations to lie in "the medium of the voice itself," in its "materiality" (Chion 1999: 1). This challenges the tendency of collapsing the voice with speech or concentrating solely on its role as the carrier of dialogue with regard to linguistic signification. Claudia Gorbman further notes in the English translation of Chion's study how "it is the voice—not as speech, not as song," but as all the qualities that remain afterward, including the voice itself "as a technological medium," that constitute Chion's subject matter (Gorbman 1999: xi).

Clearly, the qualities indicated by Chion and Gorbman coincide with the mediatized material and sensory aspects of voice. They refer to the irreducibility of such facets of voice as mode of emission (screaming, whispering), sensual impressions (evoked by intonation, timbre, recording technique, etc.), and vocally invoked spatial experiences to the voice's function as the vehicle of language and symbolic meanings. This stresses the capacities of voice to create mediatized experience in both sensorial and signifying registers. Despite their typically interlocked proceeding, these dimensions do not yield to a common identity.

A further issue highlighted by both Chion's and later discussions is that the attendance to the material, sensuous qualities of voice almost inevitably extends to the relations in which they occur and operate. Existing accounts have often attended to two types of relation. Partly overlapping, these figure between the voice and the body, and between voice and (medium-specific) visual events. As Neumark notes, the relation of voice to embodiment readily "raises the question of aesthetics" (2010: xvi). She defines aesthetics in this instance as a form of cognition achieved through the "whole corporeal sensorium" the different senses co-compose. To expand on this, the aesthetic processes that vocalities instigate in connection with particular media always already bear the marks or "writing" of their source body (2010: xvi–xvii).

The sensory materiality of vocal events and the broader multisensory arrangements within which voices function are, then, not unfamiliar concerns for investigations of voice in the media and as a medium. Moreover, the recent resurge of attention to cinema and other media as sensorial domains (e.g., Beugnet 2007)—which draws on Deleuze, Maurice Merleau-Ponty, and Jacques Rancière, among others—affords refreshed frameworks for the study of voice as sensory experience. New materialist returns to the body and experiencing, nonetheless, may contribute importantly to these developments. These approaches insist on the irreducibility of the material and sensorial threads of phenomena and experience in view of their coding within signification. Yet, they also insist that matter and sensuous experiencing are consequently not something fully unmediated or completely given at the level of their concrete actuality. By remobilizing the concepts of actual and virtual prominently developed by Deleuze and Henri Bergson, new materialist and related projects by the likes of Brian Massumi (2002), Elizabeth Grosz (2005), and others have reasserted, in contrast, the temporally and qualitatively intricate makeup of our embodied becomings and experiences in the material world.[4]

To put it concisely, virtual, here, stands for the real involvement of both past events and the as yet undetermined future in any perception that particular bodies/minds have of things in a here and now. The past enriches the attributes bodies/minds actually encounter. This takes the form of associations—subliminal and more conscious—to the "archives" of the involved perceivers' previous resonant or dissonant experiences. Most importantly and beyond any particular personalized former instance, the past conditions our very capacities to experience.

The future, meanwhile, involves the variation that experiential processes will undergo, as relations to what is sensed and experienced progress and each relational event "takes up the past differently" (Massumi 2009: 2).

These remobilizations of the notion of virtual endorse the new materialist insight according to which matter—whether it be the materiality of human sensing bodies or media imageries and soundscapes—is "always something more than 'mere' matter" (Coole and Frost 2010: 9). In excess of their spatially bounded, calculable properties, materialities involve interminable activeness and intensity of capacities to come. These actualize in their changing relations with other material entities, memories, thoughts, and so on. The vital inhering of the past in the present and the reconfiguring intensive potential of materialities, which Deleuzian-Bergsonian conceptions of the virtual offer, suggest fertile ways to explore how expressions that apparently pertain to one main sense modality, such as the auditory in the case of vocal emissions, can evoke virtual yet real *intermodal* experiences through activating our histories of perceiving vocal sounds in conjunction with such aspects as bodily gesture and visually inferable affective states. I will enquire the gains of this approach further with Cathy Berberian's *Stripsody*.

## Vocalizing the posthuman

Let us finally turn to notions about the overall ontological elasticity of voice. Be it the relations of voice to body, space, and sexual difference, or the sensorial and signifying experiences that vocal expressions inspire, previous accounts of voice in the media have regularly stressed the radical historical, contextual, and technological contingency of these characteristics of voice. Some quick illustrations must suffice here. Elsaesser and Hagener (2010: 129–33) recapitulate in their new sense-oriented approach to film theory how the technological setup of sound cinema, via separating the image track and the sound track, profoundly shattered the link between (vocal) sounds and their (human bodily) origin that may still appear obvious to us in everyday life. At the same time, this interrupted relation was subjected to diverse kinds of cinematic reconstruction.

Through examining sound-image collaborations and the myriad roles of sound transmission media in the production and dissemination of voices, media and film theoretical approaches have also interrogated what Steven Connor calls the "vocalic space." This term denotes the historically, technologically, and socioculturally mutating ways in which the voice emerges in and "procures space for itself" (Connor 2000: 12). As an example, Neumark has recently encouraged closer inspection of the relations between sound recording and modulation technologies (broadcast and other) and the development of specific embodied vocal techniques (2010: xxii).

Hence, if the vibrations of vocal sound melt in the air, research on voice/media has already shown that there is nothing solid in the aesthetics, milieus, or, indeed, human "origins" of the voice. Insofar as the voice's ontological contingency is being widely acknowledged, it can be argued that media theoretical approaches have begun expanding the study of voice beyond narrowly human-centric vistas into processes of vocalizing the posthuman. To draw on Philip Brophy, this concept refers to how the voice has, even prior to digital or earlier media apparatuses, possessed the capacity to "subsume nonhuman appellations and contort into multiple characterizations beyond itself" (2010: 361). Vocal expressions have frequently developed in relation to animal, natural, and machinic (sonic) milieus. The bodily organizations and capacities linked with voice production have also always been amenable to technical and environmental change.

New materialist approaches—now in tandem with notions of the posthuman—can again expand existing insights. As Diana Coole and Samantha Frost (2010) demonstrate, new materialist lines of thought display an ethos definable as posthuman(ist) in at least two senses. First, their ontological premise of constitutive relations replaces the elevation of the human into a higher plane supposedly distinguished by uniquely human abilities (self-awareness, meaning, culture) with the full immersion of human practices and dispositions in interrelations among material, social, human, and nonhuman factors.

Secondly, new materialist extensions of the "location and nature" of agency (Coole and Frost 2010: 9) onto technologies and other nonhuman entities pose a stimulating challenge to examinations of voice/media. In their wake, media and technologies cannot be conceived either as mere instruments for human vocal invention or as frameworks preordaining the possible forms of voice. The focus lies in how both specific media and vocalities exercise their capacities and may coevolve and individuate differently, unpredictably through their generative relatedness. I will return to this idea in the conclusions.

## *The King's Speech* and the emergent medium of the voice

Martine Beugnet is among the increasing bunch of scholars who have over the past decade reengaged the ways the material and sensory qualities of film—compositions of movement, sound, light, color, texture, and their experiencing—accompany and preempt its "construct as narrative process, system of representation, or articulation of an ideological discourse," to quote her book *Cinema and Sensation* (2007: 14). Beugnet stresses that despite its potentially forceful effects, "a sensual apprehension" of cinema is an often-backgrounded dimension of audience engagement and theory alike. This applies especially to mainstream feature films with their supposed dominance by (formulaic) plot, narrative logic, and character development (2007: 5).

Aptly, a reignited sense of cinema's sensual materiality *and* a retained sense of its narrative, representative functions are both needed to appreciate the appearances of voice and/as medium in *The King's Speech*. To begin from narrative-representational content, the relationship between voice and technical sound media is key to this film's main storyline and protagonists. *The King's Speech* presents the story of King

George VI (played by Colin Firth), who became the ruler of the British Empire in the mid-1930s after the abdication of his brother Edward VIII due to the latter's marriage to Wallis Simpson. In the beginning of the film, this focal character is still the Duke of York. The film narrates the attempts of George—or, to use his regularly evoked nickname, Bertie—to overcome his stammer. As historical source material, including audiovisual recordings, indicates, this condition characterized the actual George VI's voice productions.

The film approaches Bertie's stuttering in psychologized terms colored by popularized psychoanalytical assumptions. Not unsympathetically, his stammer is associated with his personal childhood family history, especially his relationship with his father, and consequent anxieties about assuming a public, authoritative voice. At the same time, the stammer is portrayed starkly as a disruptive exception from the vocal norm, particularly in its (male) upper class and publicly manifest guises. The need to eliminate or manage this impediment becomes increasingly pressing in the case of Bertie because of his expanding responsibilities of addressing the national audience not only at localized ceremonial and political events, but also through the radio, the then relatively new sound broadcast medium. Painfully aware of the extended spatial reach and sociopolitical weight that his vocal deliveries thereby acquire, Bertie, encouraged by his wife Elizabeth (Helena Bonham Carter), embarks on voice therapy and training sessions with speech therapist Lionel Logue (Geoffrey Rush). This is in order to rid his vocalizations of the involuntary repetitions and discontinuities comprising the symptoms of the stammer.

While centering on the evolving relationship between these two men and on a series of radio speeches Bertie prepares for, the story of the film assumes characteristics of a progress or redemption narrative. Punctuated by conventional setbacks (failed speeches after seemingly successful rehearsing, tensions between Bertie and Lionel), the movie tends toward what is envisaged as a gratifying resolution, which it at least partly reaches. By the end of the film, Bertie exhibits improved capacities of bodily and mentally controlling the sonic and signifying expressions of his voice. In the film's late scenes, these capacities are moreover pictured as transforming the kinds of affect (change of state) and ensuing experience that Bertie's vocalizations stir in the radio audiences. Growing trust in the performative nuances and nationalist messages of the King's speech become the listeners' dominant visualized states.

Quite often, the film's depictions of voice and media appear to reiterate tradition-seeped assumptions about both. For example, certain aspects of the film call forth the premise famously excavated by Derrida from the Western metaphysical tradition. According to this idea, vocal enunciations—or hearing and feeling oneself speak— epitomize the individual subject's unmediated presence to itself, or the potent illusion regarding this kind of self-derived and self-standing existence (Derrida 1998). In *The King's Speech*, this linking of voice to the subject's coinciding with itself is implied by the statement on the film's DVD case about how the movie concerns "one man's quest to find his voice," and by Bertie's exclamation "I have a voice!" at a particularly challenging moment of his vocal training. Not only does Bertie seek to conquer his stammer, then; apparently he also strives to acquire a fully fledged autonomous selfhood.

With regard to notions of medium, there are detectable signs of both determinist and instrumentalist views in the film's relationships between Bertie's deliveries and radio broadcasting. A technologically determinist view seems to surface when Bertie's father, King George V, dryly remarks in reference to the increased stress the rise of radio has put on skilled vocal performance that this medium is requiring that "we" (royal family members and other holders of public political office) "become actors." Toward the end of the film, the radio equipment and broadcasting situations become framed differently. Instead of their previous determining dominance, they now appear more like a vehicle for Bertie's changed vocal performances—as a medium partly at his command. To expand on Kember's and Zylinska's recent reflections, determinist and instrumentalist understandings of media are connected in that each tends to erect a relatively static model in which mediums, their users, and audiences preexist one another as originally separate entities. They are only brought to interaction by the intermediary layer of mediation, arguably represented in *The King's Speech* by broadcast vocalizations and speech acts (Kember and Zylinska 2012: 20–1).

Yet, I suggest that several of the film's narrative and aesthetic specificities encourage other, more minoritarian (Deleuze and Guattari 1987: 105–6) conceptions of both voice and media. These present them in emergent, constitutively relational terms. From the angle of narrative logic, Bertie's stammer may well function as an obstacle that the film's protagonists have to encounter and work through in order for them to undergo transformation and for the portrayed circumstances to reach a new state that then provides closure. Alongside its narrative positioning, however, Bertie's stuttering voice is constructed as an audiovisual event with such detail that the experiential impact of these depictions gains a partial independence from the meanings the voice is assigned on the plot level. What these audiovisual constructions cumulatively highlight is the processes of emergence and the heterogeneous intertwining factors—or instances of pre-individual relationality—that condition the individuations of a voice. The aesthetic features emphasizing emergence include close-ups of Bertie's parting and fluttering lips and visibly activated body before and during acts of enunciation—whether attempted or successful. Allied with the soundtrack, these visual qualities signal how vocal expressions necessarily and repeatedly *take* form: how they must re-arise into actuality.

Admittedly, these portrayals often seem to voyeuristically zoom into the insistent and socially stigmatizing bodily and vocal signs of Bertie's stammer. Still, because of the irreducibility of their power to the film's plot, they can draw our attention more generally to the character of voices as emergent events. All vocalizations, not just stammering ones, involve a new and partly unpredictable coming together of body, sound, subject, the surroundings onto which the vocalizations spread, and the technical media potentially involved in modulating and disseminating them. While more pronounced in some cases, this emergence, or incipience, of the voice is inherently no pathology. It is a sine qua non of vocal expressions. This realization can, in turn, inflect the understandings of the political nature of voice beyond the specter of the metaphysics of presence and any remainders of simplistic sender-medium-receiver models.

At several moments, *The King's Speech* encourages its viewers to regard the new skills that Bertie obtains in terms of broadcast voice usage and enactments of authority as desirable outcomes because they manage to reinvigorate a nationwide "listening community."[5] The film thus includes nostalgia for clearly identifiable centers of sociopolitical power and the role of media in establishing these. Nevertheless, it also gestures toward a politics of relational emergence. Comprised again of particular audiovisual details, these gestures concern both the relationship between Bertie and Lionel as well as Bertie's mediatized performances. Repeatedly, the camera follows the ways Bertie throws himself into the physical exercises Lionel introduces and enacts with him. On the soundtrack, the two men jointly try out different voice production techniques. Here, it is the moving relation between these characters and sound-making bodies with their respective pasts and inclinations that is pictured as conditioning the very actualizations of Bertie's voice. This relation enhances new vocal capacities, but it also disciplines and includes vulnerability that may diminish capacity.

In the speech scenes the camera movements between Bertie's face and body, the deployed microphones, radios, the voice therapist, sound engineers, audiences, and the speeches' sociomaterial settings implicate all these human and more-than-human factors in an interlinked capacity. These aesthetic means promote increasingly complexly the notion that the relations between the depicted activities and elements—material, social, and semiotic—form the condition for the occurring and impact of Bertie's vocalizations. The shifts of camera serve to conjoin institutional expectation with the sound-making body and its accompanying mental processes. They intertwine media apparatuses and knowledge with adjusted practices of voice training and the emergent experiences of listeners.

If it makes sense to consider the voice itself as a medium or process of mediation, in these moments of *The King's Speech* voice does not merely mediate linguistic meaning or the vocalizing body's and venue's materialities. Here we can draw on Kember's and Zylinska's recent reconceptualizations of mediation and media. Mediation, in this view, signals the primary, differing "process of media emergence" whose potentiality stems from the interplay of heterogeneous components (e.g., material, technological, and institutional). Media amount to "(ongoing) stabilizations" of this processuality (Kember and Zylinska 2012: 20–1). Following these suggestions, the processes of mediation from which Bertie's voice as a medium temporarily arises are precisely heterogeneous and complex. They encompass technological, cultural, corporeal, psychological, and additional interrelating determinants. Hence, while *The King's Speech* contains readily familiar narrative devices and conservative leanings, the ways its portrayals of voice production resonate with new materialist notions of relation and corresponding media theory enable fresher insights into the ontology and politics of voice and how these can be cinematically conveyed.

## *Stripsody* and multisensory voice

If vocal expressions emerge from relations while instigating, for their part, new relational events, then Cathy Berberian's (1925–83) vocal practice seems to have capitalized on the potential for newness and surprise (reconfiguring relationships) that the emergent character of the voice entails. Many vocal repertoires and pieces that her performances made famous might be described as true art of emergence in this regard. The extended vocal techniques to the development of which Berberian contributed are, overall, a case in point. Their premise lies in the discovery of new vocal expressions, particularly vis-à-vis the traditions of mainstream "classical" and operatic singing, through newly fashioned relations between the performer's body and such modalities of sound that mostly fall outside conventional Western definitions of musical vocalizations (e.g., speech, whispering, labial, and guttural sounds).[6]

Berberian's composition-performance *Stripsody* constitutes a singular example against this backdrop. While containing a cavalcade of elements that represent extended vocal techniques, its peculiarity pertains to what it drew its inspiration from, or the kinds of relations underpinning its emergence. As the title's amalgamation of "rhapsody" and "strip" intimates, Berberian's blueprint when composing *Stripsody* was to give an actual and exclusively vocally produced expression to a range of sounds that are implied by familiar scenarios in comic strips. These implied sonorities include yells, creaks, explosions, animal sounds (e.g., birdsong), and the acoustic signs of modern transportation technology. To build again upon Philip Brophy's propositions on the posthuman voice, Berberian's method with this piece could be described as one of summoning forth new ways in which the (classically trained) singing voice and vocal expression can "contort into multiple characterizations" beyond their more customary, expected actualizations. Clearly, the making and performances of *Stripsody* also involve the subsuming of "nonhuman appellations" (Brophy 2010: 361).

Another manner of describing the piece is that in it, Berberian endeavors to provide a soundtrack to visual presentations that typically imply dynamism: movement, energy, erupting bodily and emotive states. Although the attendant or potential sounds of these presentations are indicated visually, they do not possess an actual sonic form in the context of their original medium. However, it is not that the varying voice production styles, intonations, timbres, and dynamical shifts of *Stripsody* seek to merely sonically complement a range of medium-specific visual scenarios. Arguably, their relation to such scenarios and visual perception is more active. All but inevitably, the piece's vocals provoke actually absent, yet experientially real visual impressions as part of the perceptual processes of their listening audiences. These impressions may relate to scenes of action and emotion, movement styles, volumes and materialities of objects, bodily gestures, and comportments. This image track, if you will, may be predicated in part on the references to sound in comic book visuals that Berberian originally worked from. Still, its qualities will ultimately depend on the wider potential to hear sounds in association with optic, kinesthetic, tactile, and spatial characteristics. This potential arises from the listeners' multifarious pasts of sensation and perception.

This leads to another distinctive feature of *Stripsody*. Occasions of listening to this piece—versions by Berberian and others can be found on YouTube, for instance—are likely to demonstrate how the sensorial, experiential power of voices is not restricted to their cooperation with other sounds as well as visual (and, for example, accompanying tactile) qualities. These collaborations have already been studied in performing arts, cinema, video, other technical media, and sometimes combinations of these. Rather, the voice(s) of *Stripsody* display a vibrant capacity to mobilize the interimplication of the senses through their very sonorities alone. It is the virtual but effective persistence of past multisensory encounters in the perceptual present that enhances these instances of co-implication while lending them their evocative force. At the moment, intermodality, synesthesia, and the multisensory spectator/media user figure once again as important film and media theoretical concerns (Elsaesser and Hagener 2010: 130). However, the benefits of these discussions for reconsidering the voice in media and as a medium remain mostly unexplored. This applies particularly when it comes to interlinking these perspectives with new materialist returns to the complexity of materially based aesthetic experience at the crossroads of actual and virtual.

Berberian's *Stripsody* exemplifies how the process of mediation—or emergence—from which the voice arises as a temporary result, or medium, may include relationality between such areas that at first seem to occupy mutually distant cultural locations and address different sense channels. In this case, these comprise avant-garde singing and the (partly clichéd) expressive elements of comic book imagery. Concurrently, her vocalities illustrate how any expressive practice that is apparently linked to a particular sense modality always already activates an entanglement of the senses.

While experimenting with the specific powers of voice to implicate other senses alongside hearing, through sounds and remixes *Stripsody* invokes familiar ingredients from popular visual culture. It may help us realize our own capacities of seeing through sound. It points to the embedded nature of these capacities in our past encounters with media imageries. Moreover, *Stripsody* encourages awareness about how it is not only the signifying but also the sensorial aspects of artistic and mediatized expressions that are richly mediated by past perceptions. This virtuality of sense perception means that the form of the object—like a voice—"is the way a whole set of active, embodied, potentials appear in present experience" (Massumi 2008: 4).

## Conclusions

The impetus of this article has been to engage with and introduce new ideas to meditations on voice and media through revisiting three questions. Each appears frequently in interrogations of voice in media cultural and theoretical settings. These questions bear relevance both to voice's coalescences with specific media and to the wider literature about voice that traverses domains from film and media studies to musicology, philosophy, and cultural history. The questions concern the connective force of voices; the material and sensorial activities of voice in excess of its role as a medium for linguistic signification; and the plasticity in how the voice relates to bodies, socio-material spaces, and our notions of the human in evolving (media) cultural and technological milieus.

Unfolding with these questions, the chapter's intended contribution has sprung from new materialist theoretical propositions previously unexplored in discussions on voice and media commingling. I attempted to show how the productivity of relations in contrast to the supposedly preexisting individuality of the relating terms that new materialist accounts of matter and reality advocate promises revamped tools for voice/media studies. It expands the ways of considering the intrinsic interconnections between bodies, selves, spaces, corporeality, and technology that voices as sound events elicit. This interconnectedness has preoccupied film and media scholars, whether it comes to the portrayals of voice within particular media or its workings on auditor-spectators.

With my examples, I hoped to indicate how notions of relation as generative and constitutive may reveal fresh aspects about the media portrayals of voice (*The King's Speech*), the formation and perception of vocal expressions in contact with media (*Stripsody*), and the possibilities of conceiving the voice as medium. I also sought to exemplify how the conceptual coupling actual/virtual that new materialist and associated projects reuse facilitates refined understandings of the revived media theoretical question of sensory perception and of voice as a sensory medium.

Ultimately, the premises about the relational basis of individuation and the heterogeneity of interrelating factors propose posthuman understandings of voice and its connectedness with media. If relations are constitutive, the voice cannot be confined to exclusive or self-evidently dominating humanness in isolation from the connections its forms and affectivity entertain with technological capacity, nonhuman natural life, and other nonhuman entities. Some previous investigations have surely acknowledged the embroilment of vocal expression in such relations, but posthuman(ist) theory enables more detailed, conceptually rigorous examinations of this relationality.

Insofar as relations are ever-emergent, the voice is posthuman in a further sense: Its manifestations repeatedly exceed the by-then actualized human vocal potential. Vocal expressions to come will, with certainty—albeit unpredictably in terms of their specific quality— exceed the "currently human potential" (Massumi 2013: xxiii). Even a mainstream narrative film such as *The King's Speech* provides perspectives into how vocal sounds and a vocalizer's bodily and mental processes intrinsically interlace with sound transmission media. Concomitantly, the bond between body and voice is displayed as something that is not given but keeps emerging in dynamic interconnections with cultural, technological, and environmental factors. Meanwhile, Berberian's *Stripsody* works to expand the human vocal potential in the company of media images and nonhuman sounds. If attuned to specific instances, these new materialist reappraisals of relation, actual/virtual interplay, and the more-than-human can contribute significantly to future exploration of voices in and as media.

## Notes

1 On explorations of sound that draw from or display affinities with new materialist theorizing, see (Biddle and Thompson 2013; Tiainen 2012; Goodman 2009).
2 See also, for example, (Herzog 2010; Del Rio 2008; Sobchack 2004; Marks 2002; Marks 2000).
3 For recent remobilization of Simondon's concepts in theorizations of relationality, see (Manning 2013).
4 It should be noted that Massumi's propositions on sensation and affect as preconscious, pre-individual forces impinging on the body have drawn critique from several thinkers and directions. Especially, these critiques have called for more continuity than Massumi's approach arguably allows between previous, already signifying subjective experiences on the one hand and new sensorial, affective becomings on the other, as well as between the supposedly conscious and preconscious layers of our existence. See (Hemmings 2005; Leys 2011; Wetherell 2012: 53–67). However, particularly in his more recent texts, Massumi (2008, 2009) elaborates his take on affect, sensory experiencing, and perception by giving clear attention to the ways in which bodies experience and become affected within complex relations between their past and present encounters, milieus, and states.
5 I borrow the illustrative term "listening community" from Birdsall (2012).
6 On Berberian see (Vila 2003; Karantonis et al. 2014). On extended vocal techniques see (Potter 1998: 54–55, 170, 178).

## References

Alaimo, S., and S. Hekman (2008), *Material feminisms*, Bloomington: Indiana University Press.

Beugnet, M. (2007), *Cinema and Sensation: French Film and the art of Transgression*, Edinburgh: Edinburgh University Press.

Biddle, I., and M. Thompson, eds (2013), *Sound, Music, Affect: Theorizing Sonic Experience*, London: Bloomsbury Academic.

Birdsall, C. (2012), *Nazi Soundscapes: Sound, Technology and Urban Experience in Germany, 1933-1945*, Amsterdam: Amsterdam University Press.

Braidotti, R. (2006), *Transpositions: On Nomadic Ethics*, Cambridge: Polity.

Braidotti, R. (2013), *The Posthuman*, Cambridge: Polity.

Brophy, P. (2010), "Vocalizing the Posthuman," in N. Neumark, R. Gibson and T. van Leeuwen (eds), *Voice: Vocal Aesthetics in Digital Arts and Media*, Cambridge, MA: The MIT Press, 361–82.

Chion, M. (1999), *The Voice in Cinema*, trans. C. Gorbman, New York: Columbia University Press.

Connor, S. (2000), *Dumbstruck: A Cultural History of Ventriloquism*, Oxford: Oxford University Press.

Coole, D., and S. Frost, eds (2010), *New Materialisms: Ontology, Agency, and Politics*, Durham: Duke University Press.

DeLanda, M. (2006), *A New Philosophy of Society: Assemblage Theory and Social Complexity*, London: Continuum.

Deleuze, G. and F. Guattari (1987), *A Thousand Plateaus: Capitalism and Schizophrenia*, trans. B. Massumi, Minneapolis: University of Minnesota Press.

Del Rio, E. (2008), *Deleuze and the Cinemas of Performance: Powers of Affection*, Edinburgh: Edinburgh University Press.

Derrida, J. (1998), *Of Grammatology*, trans. G. C. Spivak. Corrected edition, Baltimore: The John Hopkins University Press.

Elsaesser, T., and M. Hagener (2010), *Film Theory: An Introduction Through the Senses*, New York: Routledge.

Goodman, S. (2009), *Sonic Warfare: Sound, Affect, and the Ecology of Fear*, Cambridge, MA: The MIT Press.

Gorbman, C. (1999), "Translator's Note," in M. Chion (ed.), *The Voice in Cinema*, New York: Columbia University Press, xi–xiii.

Grosz, E. (2005), *Time Travels: Feminism, Nature, Power*, Durham: Duke University Press.

Hemmings, C. (2005), "Invoking Affect: Cultural Theory and the Ontological Turn," *Cultural Studies*, 19 (5): 548–67.

Herzog, A. (2010), *Dreams of Difference, Songs of the Same: The Musical Moment in Film*, Minneapolis: Minnesota University Press.

Horn, E. (2007), "There Are No Media," *Grey Room*, 29 (Winter 2008): 6–13.

Karantonis, P., F. Placanica, A. Sivuoja-Kauppala and P. Verstraete, eds (2014), *Cathy Berberian: Pioneer of Contemporary Vocality*, Farnham: Ashgate.

Kember, S., and J. Zylinska (2012), *Life after New Media: Mediation as a Vital Process*, Cambridge, MA: The MIT Press.

Kirby, V. (2011), *Quantum Anthropologies: Life at Large*, Durham: Duke University Press.

Leys, R. (2011), "The Turn to Affect: A Critique," *Critical Inquiry*, 37 (3) (Spring 2011): 434–72.

Manning, E. (2013), *Always More than one: Individuation's Dance*, Durham: Duke University Press.

Marks, L. U. (2000), *The Skin of the Film: Intercultural Cinema, Embodiment, and the Senses*, Durham: Duke University Press.

Marks, L. U. (2002), *Touch: Sensuous Theory and Multisensory Media*, Minneapolis: Minnesota University Press.

Massumi, B. (2002), *Parables for the Virtual: Movement, Affect, Sensation*, Durham: Duke University Press.

Massumi, B. (2008), "The Thinking-Feeling of What Happens," *Inflexions*, 1 (1) (May 2008): 1–40.

Massumi, B. (2009), "Of Microperception and Micropolitics," *Inflexions*, 3 (October 2009): 1–20.

Massumi, B. (2013), "Prelude," in E. Manning (ed.), *Always More than one: Individuation's Dance*, Durham: Duke University Press.

Neumark, N. (2010), "Introduction: The paradox of voice," in N. Neumark, R. Gibson and T. van Leeuwen (eds), *Voice: Vocal Aesthetics in Digital arts and Media*, Cambridge, MA: The MIT Press, xv–xxxiii.

Potter, J. (1998), *Vocal Authority: Singing Style and Ideology*, Cambridge: Cambridge University Press.

Silverman, K. (1988), *The Acoustic Mirror: The Female Voice in Psychoanalysis and Cinema*, Bloomington: Indiana University Press, ix–xxiii.

Sobchack, V. (2004), *Carnal Thoughts: Embodiment and Moving Image Culture*, Los Angeles: University of California Press.

Tiainen, M. (2012), *Becoming-singer: Cartographies of Singing, Music-making and Opera*, Turku: Turku University Press.

Vila, M. C. (2003), *Cathy Berberian, cant'atrice*, Paris: Fayard.

Wetherell, M. (2012), *Affect and Emotion: A New Social Science Understanding*, London: Sage.

# 13

# Sonic Matter: The Material Cut-ups
# of Christian Marclay

### Sebastian Scherer

When we listen to music, we are usually confronted with two fundamental possibilities: we are either enjoying a live performance or we are listening to a recording, a reproduction. The live performance itself is always highly volatile and ephemeral. It is tied to musicians, instruments, the acoustic properties of the auditory space, to the audience itself, and most importantly to the moment. These variables allow for small deviations, variations, improvisations, and even mistakes that render every performance unique. And after the last note has faded it only persists in the memory of those involved. The process of listening to a recording on the other hand relies on a playback device and a fixed medium which potentially transcends the constraints of time and space: only in its material, permanent form, music can be reliably captured, retained, replicated, disseminated, and listened to repeatedly. Even despite possible shortcomings in terms of reproduction quality, every copy bears the potential to preserve and reactivate the contained music.

The material manifestations of these sound recording and reproduction media are diverse: from Edison's wax cylinders to magnetic tape playing on reel-to-reel machines, audiocassettes, eight-track cartridges, digital compact discs, and various others, especially the last century has seen (and heard) a steady succession of technological developments in this regard, yielding greater portability and sonic transparency. With every new evolutionary step toward improved fidelity the medium itself was pushed back and rendered more and more inaudible. Up to the point that it became intangible (and ephemeral) again, with the development of the Fraunhofer MP3-codec and various other digital audio standards that expand the possibilities of availability and reproducibility in a way that still poses a considerable challenge for the economic structures of the commercial music industry. However, the most iconic sound reproduction medium of the last century is without a doubt the vinyl record, a rather fragile container for the recorded music. Although the record itself is produced out of relatively inexpensive plastic,[1] it is far from being something disposable; it is often considered a highly collectible and even fetishized object, prized in enthusiast-circles. The record has many times been declared outdated and technically obsolete, but it is still not phased out, as can be seen in various musical subcultures and retro-trends that refuse to abandon the format as they praise its sonic "warmth" and the overall haptic experience. We all know the ritualistic act of playing a record, of handling the large twelve-inch cover, the protective sleeve, which keeps the heavy disc of black vinyl safe from harm, carefully placing the record on the platter of the turntable, very cautiously positioning the needle on the vulnerable surface of the record, so that we can enjoy it in its pristine condition for a long time to come.

Yet this process of playing a record also brings forward another set of listening habits, that differ greatly from the rather mobile and haphazard behavior users of digital formats display: there is no shuffle play function, no convenient skipping of tracks, no auto-repeat, no deleting of annoying songs. While it is of course also possible to listen very selectively with a turntable, we are rather more inclined to patiently enjoy one side of the record after another, flipping the disc after the first half, following the actual sequence of songs as intended by the musicians or producers of the album (Marclay and Tone 2004: 345). This way we encounter the music as a continuous stream, maybe not unlike a live performance. We may even experience the aforementioned ostensible "warmth."[2] Yet there are also other "side effects" when listening to a record: the slight hiss and crackle of the very first grooves before the music starts, the perceivable surface noise between songs, the marginal scuffing, maybe even the odd scratch making the needle skip, and of course the infinitely repeating pop of the run-out groove. While perfectionists might say that these noises detract from the actual fidelity of the reproduction, enthusiasts of course argue that they constitute an integral part of the process and rather complement the overall listening experience. From this point of view these seeming sonic deficits already hint at a vital question in regard to the materiality of the vinyl record: what creative and subversive potentials arise if we do *not* push the analog medium further into transparency, but instead welcome and even intensify these imperfections and interferences, if we willfully act against the stability of the record, if we deliberately scratch its delicate surface, and take it even beyond its material and aesthetic extremes?

This renegotiation of the materiality of the medium is essential to the early work of the Swiss-American media-artist Christian Marclay. By accepting the peculiarities and limitations of the vinyl medium as stepping stones toward an aesthetics which oscillates between music and visual art, he playfully questions conventional notions of sound and matter. Jennifer González also identifies this polar approach in Marclay's work, when she writes: "The ephemeral nature of sonic events and their material grounding in the world of objects, instruments and recording devices define the primary conceptual framework of Christian Marclay's artistic practice" (2005: 24). He thereby conflates the diverging spheres of the ephemeral live performance and recorded music (Ferguson 2003: 30), but he also transcends the static notions of reproduction and representation, making way for novel ways of thinking with media and music.

# Roots

When Marclay, after attending an art school in Geneva and studying visual arts at the Massachusetts College of Art in Boston, came to New

York in the late 1970s with a vital interest in performance art, he did not seek space and support in established art clubs, galleries, or museums (Kelly 2009: 151). The scene he was interested in met and performed in music clubs like the Mudd Club in the TriBeCa neighborhood or the legendary CBGBs on the Bowery in southern Manhattan (Licht 2003: 89). His musical and artistic socialization in the thriving New York *punk-* and *no wave-*subcultures of the late 1970s urged him to become an active contributor himself. The "anything goes"-attitude and "do it yourself"-mentality of that time and place allowed him to follow his impulse to go on stage, despite having no formal training in music at all. In an interview with Scott Macaulay, Marclay explains: "I couldn't play an instrument, so I used records. The punk movement had freed people from the idea that you had to be skilled to play music. I didn't pick up a guitar but records and turntables" (2005: 110). This egalitarian and participatory approach, rooted in the minor subcultures of performance art and punk alike, inspired a cross-fertilization of ideas that deterritorialized established hierarchies of art and music circles by playfully connecting and subverting the spheres of art and life in unforeseeable ways (O'Sullivan 2006: 74).

Marclay started to work with vinyl records in a punk duo with Kurt Henry called *The Bachelors, even.*[3] González illustrates that "the band brought experimental music and noise to the stage, smashing glass objects or chopping wood in the midst of performances. To enhance the sensory overload, the artists also projected slides, films and cartoons, and played pre-recorded tape loops to form an active visual and rhythmic backdrop to Marclay's vocals and Henry's guitar" (2005: 24). These prepared tapes often contained stuttering sounds and rhythms produced by damaged records. However, Marclay soon sensed the creative freedom and spontaneity of instead having the actual turntable and records on stage during their shows (Kelly 2009: 156). He recounts:

> The more I worked with the records the more I realized the potential
> of all the sounds generated with just a turntable and a record and
> started to appreciate all these unwanted sounds that were traditionally

rejected: skipping, clicks and pops, all this stuff that people didn't want. I started using these sounds for their musical quality and doing all kinds of aggressive, destructive stuff to the records for the purpose of creating new music. (Licht 2003: 89)

In order to not only play records on stage, but rather to perform and improvise with them, Marclay soon constructed a hybrid instrument he later named the *Phonoguitar*, "a turntable with a guitar strap, [which] allowed the artist to scratch and play an LP while holding the instrument flat to his body" (González 2005: 26). This portable playback apparatus enabled him to move on stage more freely to convey the raw punk rock energy of his band to the audience, as well as "to parody the stage performances of rock stars" (Lavigne 2007: 86). With the *Phonoguitar* Marclay developed techniques and musical gestures that stood in blatant contrast to the notion of continuous, undisturbed reproduction one normally associates with the act of playing a record. Emma Lavigne even argues that beyond that "the Phonoguitar performances reactivated what was essentially live music that was frozen in the recording of it" (Lavigne 2007: 87). Marclay's reckless manual interference with the spinning record is thus to be regarded as an act of liberation that wakes the music from its stasis in the form of grooves pressed into a plastic disc.

This breakaway from the static, representational condition of the medium was only possible because the turntable itself is technically indeed a comparatively simple device. The underlying concept of the apparatus has not changed significantly since the early days of the phonograph: there is still a needle reading the acoustic information of fine grooves engraved into a revolving disc. Later developments introduced the conversion of the needle's vibrations into an electric current, which is then equalized,[4] amplified, and translated into acoustic waves via loudspeakers. This relative simplicity and robustness of the mechanism bears great potential for manual intervention during playback. Marclay was therefore able to scratch the needle, to spin the record forward, to reverse its playback direction, to alter its pitch by slowing it down or by speeding it up, without disrupting the actual playback process altogether: Despite all disturbances induced by the artist, the record would eventually continue to revolve and the needle would find an adjacent groove to read.

Marclay started out by playing and manipulating just single records on stage, accompanying other musicians,[5] but soon he began to experiment with multiple sturdy Califone educational turntables wired to a mixing-console in order to blend just specific portions of various records playing simultaneously, thus creating dense, raw, and jagged sonic collages out of samples of appropriated sound material. These "proto-DJing" endeavors with *The Bachelors, even* have to be regarded as groundbreaking work in vinyl-manipulation, as they were developed independently from hip hop pioneers like DJ Kool Herc, Grandmaster Flash, and Afrika Bambaataa (Kahn 2003b: 60), long before the notorious "scratching" and "beat-juggling" of disc jockeys became a musical trope recognizable even in mainstream popular culture. While Marclay also perceives the turntable as an instrument of creative production, his intentions differ greatly from the protagonists of early hip hop:

> Marclay is not interested in getting people to dance, and he has very different influences from those of hip-hop, deriving from experimental practice in both music and the visual arts. ... There is no interest for Marclay in a steady beat; in fact in performance he goes to great lengths to make sure this is not possible. His scratching is not timed to accent the groove and he prepares his records so as to make certainty an impossibility. Thus despite their shared use of the turntable, the two musical practices are related to two different and divergent cultures and the two scenes have very little to do with each other. (Kelly 2009: 167–8)

Although Marclay's approach is inherently different in terms of detail and sensitivity, when compared to the manual sophistication of contemporary turntablists, Douglas Kahn asserts, that he has certainly "developed a formidable set of performance and improvisatory skills to manipulate materials and meanings" (2003b: 70).

Besides the manual interaction with the vinyl disc during performances, another method of modification devised by Marclay was to place topographic material, like small stickers or paint, onto the surface of the records to make the needle jump back to a specific point in order to create ever-repeating sonic loops. While this is also a technique nowadays frequently used by hip hop turntablists, subjecting the tender surface to such extreme measures of manipulation is normally considered a sacrilegious act that would make vinyl aficionados cringe. Kelly explains accordingly: "Marclay prepares his records, modifying them to produce additional audio through their malfunction. He scratches, cuts, and buffs his records to create an audio texture thick with rasping, ticks, and pops" (2009: 161). The continued looping and repetition of specific portions of the record would wear out the respective grooves more quickly, degrading the playback quality and rendering the captured music unstable and ephemeral again. Marclay remembers in an interview with Douglas Kahn: "I would find an interesting sound on a record and glue things, little stickers to create the loops, to make it skip. Some good loops would deteriorate quickly. I'd get depressed. … The heavy tone arms would make loops fade quickly" (Kahn 2003a: 20). The resulting loops and glitches would not only exhaust the fidelity of the vinyl, but also eventually produce barely controllable, interfering noises, obviously not existent on the original recordings. By foregrounding the materiality of the medium itself and thereby eclipsing the intentions of the performer and the original musicians, this approach yields a potential for indeterminate improvisations. Alan Licht thus identifies a relationship to John Cage's musical strategies: "In a way Marclay prepares records as Cage prepared a piano, so that recognizable sounds would be modified in an unfamiliar way. Other sounds are only familiar as accidents, but an integral part of Marclay's music" (2003: 99–100). This comparison to the prepared piano is especially appropriate, since Cage also introduced factors of indeterminacy to the instrument that extended its sonic range, but simultaneously diminished the influence of the performer and composer. Kelly also points out that "the prepared piano is a part of the trajectory of extended sound for the piano, allowing new sounds to be drawn from the instrument. In the same way the prepared record does just this; it generates sounds not originally intended by the record producer" (2009: 164).

In his performances, Marclay manipulated and juxtaposed a wide variety of records from many different musicians, composers, genres, and decades. He appropriated, manipulated, and recontextualized seemingly incongruent sound samples, which nevertheless still bear an uncanny resemblance to the original compositions and recordings they were taken from. For the audience, this encounter can be a quite amusing (or likewise alienating) experience, once some samples are identified. Kahn consequently remarks that "people in media-saturated societies have so many songs, noises, sound bites, and snippets recorded in their heads that they can register the type of quoting and referencing in Marclay's DJ work" (2003b: 70). As the outcome in most instances fundamentally differs from the source material, these resonances challenge our expectations and prevalent notions of musical unity. Marclay declares that "the idea of sampling things, collaging them, transforming them, that was a natural way of making things for my generation" (Licht 2003: 96). Kelly thus identifies him as a "quintessential postmodernist of the 1980s. His use of samples, found sounds, debris, and old technologies is a bricolage of incongruent cultural elements, which, in his hands become ripe for exploitation" (2009: 154–5). Nevertheless, such samples do not have to be readily identifiable, as Marclay considers them "fragments of our cultural baggage."[6] Even if we have not heard (or do not recognize) the particular original songs from which he samples and borrows, they still refer to a set of similar sounds, songs, and genres we have certainly already heard before (Higgs 2005: 89–90). Also Marclay maintains that "everybody grew up listening to all this stuff, and all this music is, in a way, already sampled in our heads" (Katz 1991: 7). He hereby seems to tap into our collective pop-cultural unconscious and establishes rather unpredictable connections (Higgs 2005: 90).

While nowadays most DJs seem to confine themselves willingly to the tight constraints of very specific genres and even sub-genres, Marclay took an altogether different approach: When it came to the selection of his records, he did not seem to be too particular. Instead he sampled from a multitude of artists, labels, genres, and eras without much discrimination. In his performances he incorporated "completely disparate musics: cuts from discrepant recordings, Tin Pan Alley recordings, classical music, waltzes, severely cut-up drum solos, and disco records" (Kelly 2009: 160). It was certainly not the complacent satisfaction of a distinguished musical taste he aimed to demonstrate. Miwon Kwon argues that

> while Marclay appears to share much with other samplers, re-mixers, and appropriationists of pop culture, be they musicians or visual artists, he fundamentally departs from the normative insofar as his work is not about savvy connoisseurship of mass media. ... Instead, he helps the audience recognize the aesthetic logic of cut-and-paste techniques, based on discontinuity and fragmentation. ... (2003: 125)

A decisive factor was, however, the inexpensiveness of his working material: he did most of his *crate-digging* in thrift-stores, which at the time held cheap, used records in great abundance, as the compact disc slowly began to dominate the market and people tried to get rid of their vinyl record collections, as their price was quickly deteriorating. Regarding this period Marclay states:

> I remember when my attitude towards the record changed from being this object to be respected, collected and stored for posterity, into a piece of plastic that had no more value than a coffee cup in the gutter. ... I would see records on the street, in the gutter, I would see thousands of records in thrift shops that nobody wanted, that nobody cared about. It was in some way that cultural change that allowed me to see a different attitude towards records, and I pushed in that direction, considering them as just a cheap commodity to be used and abused. (Marclay and Tone 2004: 345)

This pragmatic, disenchanted perspective on the record allowed for spontaneous, radical performances, illustrated by pieces like Marclay's 1985 *Jukebox Capriccio*, released on his retrospective compilation *Records 1981-1989* (Marclay 1997). González describes the works included on this CD as offering "a unique trajectory for the listener who is invited to traverse unpredictable intersections of auditory reference" (2005: 33). Licht similarly suggests that "the key characteristic of the music is the juxtaposition of kitschy jazz, exotica, dance music, and classical sounds, often looped or warped by changing turntable speeds and scratching, with the records' actual surface noise always present to remind the listener of the vinyl as the sound source" (2003: 99). The title of the track *Jukebox Capriccio* already gives it away: it is a freeform musical piece, which playfully violates musical rules and conventions, by using stuttering off-beat rhythms, repeating loops of out-of-key notes, piercing string-sections, bursts of scratches, and distorted noises produced by the playback devices and the abused records. While the certainly most recognizable musical reference in *Jukebox Capriccio* is the intro from Soft Cell's 1981 cover version of Edd Cobb's *Tainted Love*, Licht laconically explains: "'Jukebox Capriccio' has a big band overlaid with classical orchestra recordings spinning at hyperspeed" (2003: 100). Marclay's approach in this piece surely evokes humorous moments, but the overall impression remains rather disorienting and unsettling, due to the multitude and fast succession of sonic sources and effects he combines. In addition, *Jukebox Capriccio* also bears a certain subversive momentum, as its juxtapositions unveil recurring musical patterns and stereotypes, like dramatic glissandi, exhilarating drum-rolls, or suspenseful brass entries.

## Influences

Nowadays we can rely on various taxonomic categories to approximate Marclay's compositional strategies in this piece: it could, for instance, be described as an improvised live *remix* or an anarchic, analog *mash-up*. However, when *Jukebox Capriccio* was created, these pop-cultural forms of musical bricolage were still just in their infancy. The commonly required technology, like the digital samplers, would not become more affordable and accepted among a wider range of musicians and producers until the late 1980s. In this sense, Marclay's work can be regarded as a crude analog precursor to what would later evolve into variety of contemporary experimental music genres, centered on refined sampling methods. Nevertheless, it should not be ignored that Marclay's approach is also indebted to certain stylistic influences and conceptual predecessors in music as well as in art history.

In the early twentieth century, the American modernist composer Charles Ives also combined musical fragments and quotations from a variety of different sources. For his own compositions he borrowed portions of hymns, classical music, and traditional songs, referring the listener to the original pieces, albeit not by the detour of using recorded audio material, but instead by amalgamating his sources already during the writing of the musical score.

A certain similarity to indeterminate pieces like John Cage's *Imaginary Landscapes* is also apparent. Yet "while Cage's Imaginary Landscape No. 1 incorporated records (and a radio) as early as 1939, his influence on Marclay is perhaps more broadly evident in his embrace of music as a part of a wider field of audible experience" (Ferguson 2003: 41). This is underlined by Cage's impartiality in regard to the use of noise in his compositions. In his manifesto "The Future of Music: Credo" he wrote as early as 1937: "I believe that the use of noise to make music will continue and increase until we reach a music, produced through the aid of electrical instruments which will make available for musical purposes any and all sounds that can be heard" (Cage 2011: 3–4). In this regard, Marclay's indiscriminate work with electrical turntables and a wide variety of records can indeed be comprehended as a postmodern extrapolation of Cage's ideas.

Also the tape experiments of *Musique Concrète* come to mind here. The implicitness with which composers like Pierre Schaeffer used technical equipment to record, manipulate, and combine the sounds of everyday life, suggests a certain conceptual relation. Yet Licht identifies a crucial distinction which situates Marclay in an altogether different context: "Creating sound collages with thrift-store records and Califone turntables as opposed to expensive reel-to-reel tape in an expensive or government-funded studio, as *Musique Concrète* composers like Pierre Schaeffer and Pierre Henry did, is in line with the punk ideology" (2003: 91).

Marclay's collages not only exhibit a certain kinship to the acoustic anarchy of punk music; they could even be understood as sonic counterparts to the cut-and-paste aesthetic of early punk rock flyers, magazines, and album covers. An iconic example for this visual approach is Jamie Reid's provocative cover-artwork and promotional poster for the Sex Pistols' 1977 single *God Save the Queen*, which is composed out of a crumpled Union Jack flag, superimposed with a black-and-white photograph of Queen Elizabeth II and newspaper clippings covering her torn out eyes and mouth, giving it the jolty and scandalous appearance of a ransom note.

Aesthetically related methods of fragmenting and processing appropriated material also appeared in the visual arts throughout the twentieth century: in the European dadaist circles of the 1910s and 1920s we already encounter the eclectic techniques of montage and collage. For instance in Hannah Höch's 1919 *Schnitt mit dem Küchenmesser Dada durch die letzte weimarer Bierbauchkulturepoche Deutschlands,*[7] which features an onslaught of combined magazine and newspaper clippings, depicting grotesque scenes of the aftermath of the First World War militarism, industrialism, and the social upheaval of that time. The process of selecting copies, of cutting them apart and pasting them back together is essential here. The resulting jagged, kaleidoscopic multiplicities could be read as aesthetic reactions to the alienating, fragmented experience of fast-paced inner city life back in the early twentieth century, which defied comprehension of the world in its totality.

In this context of dadaism it was Marcel Duchamp who pioneered the groundbreaking idea of the readymade, which proved the point that a finalized, appropriated object, even an industrially produced commodity, could stand as a work of art in its own right, once it is chosen, named, and signed by the artist (Ferguson 2003: 40). One of the most notorious examples for this approach is of course his 1917 artwork *Fountain*, a tilted and signed porcelain urinal, which questions the importance and integrity of artist, artwork, and audience alike.

In the 1950s and 1960s the protagonists of pop-art similarly relied on the creative appropriation and reproduction of the iconography of commercial advertisements, packaging, and objects of popular culture. Andy Warhol's 1962 *Marilyn Dyptich*, which features the artist's famous silkscreen permutations of a press photo of Marilyn Monroe, serves as a prime example to illustrate this scrutinizing of consumer society and material culture, by means of relentless reproduction, recombination, and deterioration. In this context, the resulting aesthetic could also be associated with the *cut-up* and *fold-in* techniques, pioneered by experimental writers of the Beat Generation, like Brion Gysin and William S. Burroughs, who rearranged fragments of found texts and newspaper clippings in order to devise a new literary experience.

Also, the Fluxus artists of the 1960s and 1970s followed a similar approach influenced by the ideas of John Cage, when they incorporated found objects and simple acts of everyday life into their *neo-Dadaist* performances. Their focus was not on the gradual perfection and unity of a finalized artwork, but rather on the process itself, which playfully questioned the established dichotomy of *high* and *low* culture, of art and the mundane activities of everyday life (Kelly 2009: 152). Especially their provocative performances that involved the destruction of musical instruments in order to produce new, indeterminate sound can be related to Marclay's own artistic strategies.[8] In this context the Czech Fluxus artist Milan Knížák is also worth mentioning, as he already tried to test the acoustic and aesthetic possibilities of manipulating vinyl records in the 1960s with his *Broken Music* experiments. However, Kelly identifies fundamental differences to Marclay's work:

> Milan Knížák did play his reformed records, but this was carried out simply by allowing the needle to play a single work at time. This is very different from the manner in which Marclay performs his discs, utilizing multiple turntables and multiple record works to create a

performance event. Another difference between the two is in the playful nature and lightness of the sounds used by Marclay. Knížák's audio is thick with serious sounding classical music while Marclay's sounds come across as humorous and light-hearted. (2009: 162)

Regarding this multitude of influences, Kwon claims that "loosely interpreting various artistic strategies inherited from Cubism, Dada, and Surrealism, as channeled through Pop, Fluxus, and punk rock, [Marclay] has created music, objects, images, and performances through sampling, appropriation, collage, and montage—all related techniques of production based on radically redirected modes of consumption" (2003: 121–2). As his raw material, the vinyl record, is in this sense also to be regarded as an industrially produced and finalized consumer good, Marclay's artistic practice can indeed be thought of as based on appropriated readymades. Ferguson agrees that "one of Marclay's breakthroughs was his realization that recorded music could be treated as a form of readymade [although] he has never made what one might call 'pure' readymades. While he certainly appropriates [he] has always manipulated the material with which he chooses to work" (2003: 40).

## Recycled records

A next, even more radical, extension of Marclay's artistic manipulations was to transfer his sonic collage method to the record itself. He began to perceive the medium, the mere material container of sound, as something he could sculpturally work with. Instead of just modifying its surface, Marclay started to literally cut up the vinyl records and to paste the rearranged fragments back together, thus creating a physical remix of a variety of musical sources, highlighting the material properties of the manipulated vinyl as well as the sonic properties of its original content. Concerning his source material, Marclay accordingly stresses: "Records are … fragile, so we respect them and use them carefully. … I destroy,

I scratch, I act against the fragility of the record in order to free the music from its captivity" (2005: 115).

Depending on the ambient temperature, vinyl is a relatively flexible and soft synthetic compound: we all know how easy it is to accidentally scratch it. Vinyl can be cut, but only with great diligence and the right tools. Once it breaks, however, it shatters to sharp and jagged fragments. As can be seen in various examples, Marclay used both methods—cutting and breaking—to disassemble his records. To join the slices of various different discs back together again, he simply reglued them. The cuts and breaks in the *Recycled Records* differ from those in Marclay's earlier performances as they do not only harm the record's surface, but they disrupt the entire vinyl disc. However, these fractures allow for a degree of haptic and sonic immediacy in recombining the shards, unmatched by more elaborate, inconspicuous, nondestructive forms of editing. Deliberately lacerating and deconstructing the delicate surface of the vinyl again implicates the breaking of a taboo, as most consumers consider the record as something very precious, a fragile item to be handled with extreme care. Marclay on the other hand seemed to have a rather unemotional, pragmatic relationship to his working material:

> I don't necessarily like a lot of the records that I use. My association with them is not through a preference for a certain record; it's more a dislike. The reason I use those records is that they're easily available, they're cheap, they're just junk. It's through what has been discarded, what people don't like anymore, that I like to create something different. (2005: 124–5)

With this approach Marclay was able to create unique works from disposable, mass-produced copies of commercial popular culture.

The visual results of these eclectic recontextualizations appear at times crude and improvised, yet in other pieces very geometric and accurately executed. By using colored vinyl and picture-discs for his montages, Marclay was able to develop intricate patterns, shapes, and even figurative elements within the confines of his circular twelve-inch "vinyl-canvas." Ferguson argues: "These cut-up and reassembled records could still (just) be played. Their visual qualities, however, increasingly disrupted their capacity to deliver any coherent music" (2003: 21). In some of Marclay's works it is the visual aspect that dictates the sonic outcome—in others a reversal of this relationship can be observed. In both cases they still allude to, on the visual as well as on the aural plane, the heterogeneous pieces they were made of: "His Recycled Records ... are not neutral shards of a collage but the visual notation of stutter-step rhythms and curt segues among different musics, musicians, and their attendant social scenes" (Kahn 2003b: 70). The *Recycled Records* are hybrids between musical compositions and visual art-objects, which can be exhibited in a gallery or a museum, but also played at a concert or a performance. Marclay remembers:

> The only objects I was making then were these homemade records, The Recycled Records I used in performances or displayed on turntables so people could listen to them. I liked the idea of making a record that was unique; every time you play it it plays differently, the needle jumps around. (Licht 2003: 90)

By subjecting the record to this seemingly catastrophic event, Marclay opened yet another scope of sonic possibilities that goes beyond what he did in his performances with multiple turntables. By fragmenting, altering, and suturing the record (and thereby inevitably dissecting the contained music) he emphasized the materiality of the record itself and shifted the focus of the listener to the sounds that are usually deemed as unwanted: the crackling and skipping noise of the needle are elevated to the status of an aesthetic device in their own right: "The surface noise, the pops, the crackles, all those unwanted sounds, occur with time. I don't reject them, I use them to my benefit, value them and enjoy them. This sound patina has a great expressive quality" (Marclay 2005: 121).

Essential to Marclay's artistic practice is not only the disintegration, but also the reconstruction of the record. The title of the *Recycled Records* series is an obvious allusion to recycling, in terms of putting waste material to new use. Yet Marclay also *recycles* the records, he literally makes them round again, by means of artistic appropriation, dissection, and splicing of the material. However, a sense of unity and coherence does not emerge. The resulting noises, scratches, and cuts in Marclay's *Recycled Records* rather echo the aesthetic experience of fragmentation. They sabotage the conception of an undisturbed listening experience and remind the listener of the pretense a recording constitutes. Throughout the last century, the fidelity and transparency of sound reproduction media has been gradually refined in order to detract the listener's attention from the sonic limitations of these media and playback devices and rather emphasize the recorded performance of the original artist. Regarding his contrary intentions, Marclay consequently states:

> With music, I want to disrupt our listening habits. When a record skips or pops or we hear the surface noise, we try very hard to make an abstraction of it so it doesn't disrupt the musical flow. I try to make people aware of these imperfections and accept them as music. (Estep 2001: 80)

Like abstract painters shunning the representational tradition of their predecessors, who in turn strived for a painting to depict reality as accurately as possible, Marclay acknowledges that the actual music on a record is not reality itself, but a mere reproduction of it—a representation. Moreover, advanced recording- and studio technology regularly produce music that actually sounds *larger than life* and has very little in common with the captured live performances of musicians. Transparent recording media and high-fidelity reproduction go hand in hand: they merely aim at creating an illusion for the listener. With the advent of stereo-sound, the layering of multitrack recordings, overdubbing, and a multitude of different audio-effects, recorded music for the first time had the potential to take on hyperreal, or even surreal, forms:

> By the early 1960s … recording had already begun to separate itself from the direct trace of performance. Phil Spector's "wall of sound" … was

unequivocally a studio assemblage. ... The production of a recording became more like a composition, more a work of art in itself than a transcription of one. (Ferguson 2003: 30)

In this pursuit of the perfect illusory recording take, the traces of the actual recording process and the inherent sound of the recording equipment and media are commonly minimized or edited out by all technical means. Marclay on the other hand does not try to cover the tracks of his working process; instead, he forcefully renders them perceivable. In an interview with Michael Snow he asserts: "... Awareness and suspicion of the illusion and the fictional are important to me" (2007: 129). Kwon consequently argues:

What distinguishes Marclay's work from others who recombine fragments of found cultural material ... is the extent to which the coming together of the parts remains intelligible as a structurally integral aspect of the work. That is, the "seams" are not smoothed over to create the illusion of a "natural" whole. Meaning lies in the seams. (2003: 123)

Marclay's crude "seams" and indeterminate cuts subvert the illusion of musical continuity inherent in the studio recording. Kelly claims that "Marclay, however, hears the pops and ticks, especially on old and discarded record discs, as material for his music and questions the desire for perfect sound reproduction" (2009: 172). These microscopic failures and ruptures of the medium do not imitate the recorded sound of a real instrument or a performance, but produce an unforeseeable, abstract aesthetics, not intended by those involved in the production of the original recording. They incessantly evoke the materiality and decay of the medium and let the musical illusion deteriorate. Ferguson also affirms that "the pops and scratches that punctuate Marclay's music are the audible trace of records as real-world artifacts rather than as transparent media" (2003: 41). Marclay puts it in even simpler terms: "The recording is a sort of illusion while the scratch on the record is more real" (Estep 2001: 80).

By welcoming these transient, nonrepresentational sounds of the manipulated record and the skipping needle, Marclay's approach also alters and overrides the predetermined, linear sequence of the playback process. The needle is thrown off course and immediately picks up another contingent groove. Kelly explains: "The join in these Recycled Records cannot be perfect, causing a loud click as the needle hits the glued break. As the grooves do not quite line up, the reaction of the needle to the cut is extremely unpredictable" (2009: 161). These repeated, deliberate breaks and sutures impose their own complex rhythmical structures on the composition, dictated by the variable speed of the revolving turntable platter. They open Marclay's work to the categories of indeterminacy and chance, ultimately emphasizing not the artist, not the music, but the medium itself. Yet the glitches are not merely instances where the record fails and the music is eclipsed by a brief burst of noise. They rather mark passageways that penetrate sound recording history in nonlinear, indeterminate ways, like wormholes that fold the topology of the musical continuum, instantly taking the listener to remote acoustic places, that do not necessarily confirm expectations, but still sound strangely familiar. Marclay's dynamic, referential multiplicities might evoke moments of recognition or even amusement, but they ultimately cease to represent. They can be understood in a context that Drew Hemment identifies as a "counter-history marked by accident, manipulation and reuse that detached itself from the *telos* of representational technologies" (2004: 80). Marclay's *Recycled Records* might resonate with our musical knowledge, yet their ever-changing, nonhierarchical configuration denies any emergence of coherence and deeper meaning and renders them just as abstract as the glitches themselves.

These transient noises that occur during Marclay's performances relate his work to chance composition in a Cagean sense. But while John Cage regularly aimed for a radical minimization of the composer's influence and a subsequent depersonalization of the music, Marclay's approach is more subjective and controllable. After all, he selects the records, prepares them, and makes subjective decisions during his performances. Yet the outcome is certainly indeterminate as the skipping needle allows artistic control only up to a certain degree. Ferguson concurs: "The attraction to improvisation and chance, always in tandem with a parallel impulse towards precise control, is … characteristic for Marclay's work …" (2003: 33). Kahn thus claims that Marclay's music demands for a more "complex listening" in order to "grasp the multiple foci of musical and auditory phenomena in which many different sounds can occupy the same space and time" (2003b: 63). The rapid alternations of kaleidoscopic musical references and indeterminate disruptions produced by the rotating vinyl shards are certainly not light fare: the continuous omission and substitution of parts of the original recordings evokes a fragmented, oscillating experience, not only between sculpture and sound, but also between presence and absence, structured by the contingent rhythms of the breaks and cuts as well as the mesmerizing optical patterns and spirals of the gyrating *Recycled Records*. Ferguson also agrees that Marclay's "music demands an engaged listener, one willing to accept the challenge posed by the barrage of sound that constitutes one of his performances" (2003: 42). But how could we delineate such a more complex, engaged listening? Does it have to follow a mode of perception and interpretation that produces a linear, coherent, narrative of recognition and meaning? Is it not rather an indeterminate, ephemeral thought process that the encounter with Marclay's musical novelties provokes? Music can indeed be perceived as "an opening for thought" (Hasty 2010: 1). But Marclay's work does not foreground the mere representational capacities of music; instead it rather invites us to a thinking informed by the material manifestations of music and the indeterminacies of its own generation. Murphy and Smith suggest in their essay "What I hear is thinking, too" that Deleuze and Guattari's thinking in assemblages approximates the particularities and complexities of musical approaches that play with and subvert the logics of pop. Even pop itself "can be conceived as a rhizome because it develops by fits and starts, in a messy, practical, improvisational way rather than in a refined, programmatic, theoretical way" (Murphy and Smith 2001).

## Pop assemblages

Whether it is one of his performances with manipulated records or his sculptural *Recycled Records,* it is the process that is essential to Marclay's art. His musical multiplicities trace an unstable flux, a becoming that exceeds the limitations of its appropriated source material. They maintain an openness and connectivity that bears the potential for further improvisational proliferation. In this way, Marclay's approach relates to what Deleuze describes in his *Dialogues* with Claire Parnet: "Some contemporary musicians have pushed to the limit the practical idea of an immanent plane which no longer has a hidden principle of organization, but where the process must be heard no less than what comes out of it" (Deleuze and Parnet 1987: 94). This processuality is manifested not only in Marclay's improvised referencing and transforming of appropriated material—it becomes especially audible (and visible) in the traces, the scratches, the breaks that commemorate the artist's working material. These breaks indicate lines of flight that yield Marclay's musical de- and re-territorializations. His eclectic method extends in unexpected diagonals that conjoin and juxtapose remote sonic fragments. Edward Campbell suggests that "[the line of flight] forms itself between already existing points and lines and doing so creates a new line in a new space. This new line, this diagonal or transversal, marks out a philosophical or musical territory of its own, one which has never been known before" (2013: 40).

Thus, Marclay's samples are not only deterritorialized from the unity of their former compositions, genres, scenes, but then again reterritorialized in his novel performances and assemblages, often in a rather subversive manner. His rampant approach, influenced by the minor subculture of punk, defies conventions and economic structures of the mainstream music industry, whose products are ironically the sources of his samples. Murphy and Smith argue that "music is deterritorializing when it moves in a different direction, that is, when it no longer gives primacy to formal relations and structures, but to the sonorous material itself" (Murphy and Smith 2001). The breaks in Marclay's material entail fragmented experiences that challenge the presumed coherence and unity of a record. They can be understood as asignifying ruptures, as proposed in Deleuze and Guattari's rhizome model: "A rhizome may be broken, shattered at a given spot, but it will start up again on one of its old lines, or on new lines" (1987: 9). Like Marclay generates an intricate network of "new lines," of plurivalent cultural references, the rhizome proliferates in a ramified structure, which exceeds arborescent notions of linearity and hierarchy, and facilitates the thinking of heterogeneous, decentered, and indeterminate multiplicities. Similar to Marclay's work, rhizomatic thought is characterized by principles of connection and heterogeneity, which imply that all points within a rhizome can, and eventually will be connected to each other, no matter how different they are. The rhizome has multiple possible entries and exits and it is unpredictable in its movements, it keeps expanding, and is open to change (1987: 21). In the same way Kwon argues that "Marclay insinuates the aesthetics of collage as a social and political paradigm, one that embraces heterogeneity, the accidental, the disjointed, the forever incomplete" (2003: 127). Deleuze and Guattari confirm the relevance of the rhizome model for music, as it "has always sent out lines of flight, like so many 'transformational multiplicities,' even overturning the very codes that structure or arborify it; that is why musical form, right down to its ruptures and proliferations, is comparable to a weed, a rhizome" (1987: 11–12). The nonlinearities and discontinuities in Marclay's artistic interventions indeed overturn the codes, conventions, and genealogies of popular music.

In this sense, the rhizomatic multiplicities and asignifying ruptures constituted by the artist, his turntables, his appropriated records, as well as his audiences, can be understood as emerging assemblages (1987: 8). Campbell explains: "All creative innovation, in fact, is said to involve processes of de-territorialization in which concepts break down and are uprooted from their context only to reassemble with other heterogeneous elements to form new assemblages" (2013: 39).

Marclay's assemblages indeed counter and uproot our listening habits by disrupting the presumed linear succession of musical time and by exposing the illusion of transparent reproduction. He brings together "different dates and speeds" (Deleuze and Guattari 1987: 3), artists, styles, and genres, without subjecting his works to the regimes of unity and representation. Instead he subverts the materiality of the medium to bridge the divide between the ephemeral live performance and static reproduction. At this pivotal point Marclay probes and investigates the inner mechanics and materials of popular culture and music. He thus reflects on our relationship to the medium, by conceiving it not as a finite, fixed entity, but rather as something that invites experimentation, a perpetual transformation of heterogeneous sonic glimpses, a becoming, which defies unity, representation, and ultimately interpretation. Hemment calls this "a new kind of music making, one based in a foregrounding of interference, citation and secondary processes, a plastic art working within and through the grain of the machine" (2004: 80). Deleuze and Guattari condense it even further when they state: "RHIZOMATICS = POP ANALYSIS" (1987: 24).

Yet, despite all the "detritus of mass culture" (Cox 2010: 8) and despite all the identifiable art historical and musical lineages compiled in Marclay's work, it is futile to desire comprehension and emergent meaning. Instead, the ruptures, breaks, and disenchanting noises that continually revolve around the center spindle of the turntable—an axis that suggests infinite cyclical repetition—sing an abstract refrain that gets center stage. Salomé Voegelin hence describes her encounter with Marclay's music as follows: "I fall into the sensorial rhythm of the work and abandon any attempt to summarize or judge it. It is there to be heard" (2010: 61). Similarly, Deleuze claims that "there's no question of difficulty or understanding: concepts are exactly like sounds, colours or images, they are intensities which suit you or not, which are acceptable. Pop Philosophy. There's nothing to understand, nothing to interpret" (Deleuze and Parnet 1987: 3–4).

# Notes

1  Polyvinyl chloride, PVC.
2  What vinyl enthusiasts refer to as "warmth" is caused by nonlinearities in the frequency response curve of the medium and its playback devices.
3  The band was named after Marcel Duchamp's 1923 artwork *The Bride stripped bare by her bachelors, even,* which already hints at the reverberations of a certain dadaistic sensibility in many of Marclay's later artistic endeavors. (Ferguson 2003: 33)
4  The equalization is implemented according to a frequency response curve specified by the Recording Industry Association of America, in order to minimize hiss and noise, and to extend the maximum recording time of a record.
5  Among them were the acclaimed avant-garde saxophonist John Zorn and the guitarist Elliot Sharp. (González 2005: 27)
6  Although Marclay refers here to his video-installation *Video Quartet* (2002), it is fair to argue that the term "cultural baggage" is also appropriate in the context of his work with vinyl records, as the mode of appropriating and recontextualizing his source material is similar. (Higgs 2005: 89)
7  "Cut with the Dada Kitchen Knife through the Last Weimar Beer-Belly Cultural Epoch in Germany."
8  At the 1962 FLUXFEST in Wiesbaden Philip Corner, Dick Higgins, Emmett Williams, Nam June Paik, Benjamin Patterson, Wolf Vostell, and George Maciunas celebrated the physical destruction of a grand piano in a performance called "Piano Activities."

# References

Cage, J. (2011), *Silence*, Middletown: Wesleyan University Press.
Campbell, E. (2013), *Music after Deleuze*, London and New York: Bloomsbury.
Cox, C. (2010), "The Breaks," in C. Barliant (ed.), *Christian Marclay: Festival. Issue* 3, 6–15, New Haven: Yale University Press.
Deleuze, G., and C. Parnet (1987a), *Dialogues*, transl. H. Tomlinson and B. Habberjam, New York: Columbia University Press.

Deleuze, G., and F. Guattari (1987b), *A Thousand Plateaus. Capitalism and Schizophrenia*, transl. B. Massumi, Minneapolis and London: University of Minnesota Press.

Estep, J. (2001), "Words and Music: Interview with Christian Marclay," *New Art Examiner*, 29(1) (September–October 2001): 78–83.

Ferguson, R. (2003), "The Variety of Din," in R. Ferguson (ed.), *Christian Marclay*, 19–58, Los Angeles: UCLA Hammer Museum, Steidl.

González, J. (2005), "Survey," in J. Gonzáles, K. Gordon and M. Higgs (eds), *Christian Marclay*, 22–82, London: Phaidon.

Hasty, C. (2010), "The Image of Thought and Ideas of Music," in B. Hulse and N. Nesbitt (eds), *Sounding the Virtual: Gilles Deleuze and the Theory and Philosophy of Music*, 1–22, Farnham and Burlington: Ashgate.

Hemment, D. (2004), "Affect and Individuation in Popular Electronic Music," in I. Buchanan and M. Swiboda (eds), *Deleuze and Music*, 79–94, Edinburgh: Edinburgh University Press.

Higgs, M. (2005), "Focus," in J. Gonzáles, K. Gordon and M. Higgs (eds), *Christian Marclay*, 83–91, London: Phaidon.

Kahn, D. (2003a), "Christian Marclay's Early Years: An Interview," *Leonardo Music Journal*, 13: 17–21.

Kahn, D. (2003b), "Surround Sound," in R. Ferguson (ed.), *Christian Marclay*, 59–81, Los Angeles: UCLA Hammer Museum, Steidl.

Katz, V. (1991), "Interview with Christian Marclay," *The Print Collector's Newsletter*, 22(1) (March–April): 5–9.

Kelly, C. (2009), *Cracked Media. The Sound of Malfunction*, Cambridge and London, MA: The MIT Press.

Kwon, M. (2003), "Silence Is a Rhythm, Too," in R. Ferguson (ed.), *Christian Marclay*, 110–27, Los Angeles: UCLA Hammer Museum, Steidl.

Lavigne, E. (2007), "A Walk on the Wild Side," in J. Criqui (ed.), *Replay. Christian Marclay*, 80–94, Zürich: JRP Ringier.

Licht, A. (2003), "CBGB as Imaginary Landscape. The Music of Christian Marclay," in R. Ferguson (ed.), *Christian Marclay*, 88–103, Los Angeles: UCLA Hammer Museum, Steidl.

Marclay, C. (1997), *Records 1981-1989*, Chicago: Atavistic Records.

Marclay, C. (2005), "Artist's Writings," in J. Gonzáles, K. Gordon and M. Higgs (eds), *Christian Marclay*, 106–44, London: Phaidon.

Marclay, C., and Y. Tone (2004), "Record, CD, Analog, Digital," in C. Cox and D. Warner (eds), *Audio Culture. Readings in Modern Music*, 341–7, New York and London: Continuum.

Murphy, T., and D. Smith (2001), "What I hear is thinking, too," *ECHO*, 3(1). Available online: http://www.echo.ucla.edu/Volume3-Issue1/smithmurphy/smithmurphy1.html (accessed August 3, 2014).

O'Sullivan, S. (2006), *Art Encounters Deleuze and Guattari. Thought Beyond Representation*, Basingstoke and New York: Palgrave Macmillan.

Snow, M. (2007), "Michael Snow and Christian Marclay: A Conversation," in J. Criqui (ed.), *Replay. Christian Marclay*, 126–34, Zürich: JRP Ringier.

Voegelin, S. (2010), *Listening to Noise and Silence*, New York and London: Continuum.

# 14

# Media Disenchantments

Thomas Köner

As a media artist, I am combining visual and auditory experiences. I am interested in the relation of image and sound. Knowing that no such relation exists—two separate senses are addressed that function independently—does not make it easier. Why this combination? Is combining the flavor of certain spices with major triads less arbitrary? Or accompanying pressure on someone's sole of foot and left shoulder with a gradient from yellow to green? While smelling dandruff? As it is, to me media matter means some kind of (expanded) image, and the audio helps selling it, be it in a pavilion art installation or perfume commercial: it is a dirty job, and somebody has to do it. I am going to talk about four works of mine that serve as examples for my use of media matter.

*Banlieue du Vide* is an audiovisual piece that I did in 2004 (Köner 2004). Back then, I did not have a camera. But I had access to a Finnish traffic surveillance website. I was especially interested in pictures of roads without traffic, and during two or three months I was up at night to grab pictures of deserted Finnish roads and intersections. I collected a sufficient amount of frames to construct a video. Later, on my way to a meeting, I had to stop on a pedestrian crossing in the middle of a road in Cologne. The sound was superb, spacious, symphonic, and I recorded it. This recording became the material for the sound in *Banlieue du Vide*. The material underwent massive manipulations. I crushed the audio, it became a grainy grisaille of traffic, I left the images untouched, and manipulated the time. The frames had been collected during the course of two or three months but were compressed to the ten-minute duration of the piece. Yet, as a video, it appears unnaturally slow, close to a cinematic standstill. It was important to me that the piece is somehow boring. Boring is good for me, it sharpens the attention, helps to stretch the attention span, and even prolongs my life: ten minutes that feel like an hour are fifty minutes of lifetime for free. I believe that being able to observe or understand the totality of processes in the universe would be the ultimate boredom, so boredom to me contains the promise of truth. And the sources for this piece were rather random, starting with a Finnish friend showing me a boring website.

The next example is the performance piece *Alchemie* with filmmaker Jurgen Reble (Köner and Reble 1992). In this performance Reble is developing and treating with chemicals a five-meter film loop while projecting it with a sixteen-millimeter film projector. This creates unseen and unexpected images, especially if he starts with black film and just scratches it, while the film loop is moving through a variety of cups with chemical liquids, and the occasional puff of smoke. To create a sound for this you would probably not take your guitar and start singing. I positioned microphones near the cups, capturing some steam sound and the gurgling of the liquids, and I used pick-up microphones inside the projector itself, the shutter, the fan, the transport, and the motor. All these signals are fed into a mixing board, a bouquet of dramatic sounds that are amplified with a big speaker system. This adds to the dramatic impression: the film will "die" after forty-five minutes when the carrier material is decomposed and the lamp burns the first frame and destroys the loop.

Working with whatever media material that is already there means I have less choice. In fact, in my experience, it is true that I have no choice at all. Why? Because you play a role, that of an academic, and I play a role, that of an artist. We call it choice because it sounds nicer— our constant rearranging to always willingly be accustomed to the normative requirements we are subjected to, totally determined. It is pure ideology to say that I am in charge of my biographic narrative, while everything, including even the impression of free will, is determined by circumstances, which itself depend on a complexity of circumstances that probably involve all particles in the universe. Not only any choice, but also the given material is a trap, a prison, a lie. The image is untrue. A fixed snapshot of a process, of something that is never still, is a snapshot of something that does not exist. It fuels the ideology of the point of view. But there are infinite points of view, and the picture frame is concealing literally everything. Pitch is a prison. Duration is a prison. So I write sound color compositions instead that avoid melody and rhythm. Of course, the gesture of rejecting the prison is itself a prison, as the gesture is always already contained in and pre-formed by the rejected matter.

According to my experiences, there is no escape of the illusion. Splinters of everyday occurrences provide the material for my works. I had a video camera for a while. The REC button was painful as if it was charged with high voltage. I pressed the button *n'importe quoi*, for five seconds, wherever. Every scene is material, and equally appropriate.

In the next example, *Peripheriques pt3: Buenos Aires* I videotaped children for five seconds on their way from school in a shantytown of Buenos Aires (Köner 2005). I manipulated the image, what you see now are empty white coats moving/floating on a dirt road, no hands, no faces, no bodies, just empty coats. Obviously I am an artist, an entertainer. I started playing the violin at the age of four. I remember our cat running away in panic when I came close to my violin case. The conductor of my youth orchestra was always wearing white polyester polo-necks and I stared at the sweat stains in his armpits. I was scared to play the wrong notes and trying not to lose track in those bleak scores, as a second violin has nothing else to play than confusingly long notes and pauses. The violin maker who built the first violin was probably also scared, trying to find a tone that could please the Monarch. Still everybody is trying to please whatever Monarch. In case of a violin, it is built-in, just as in any other instrument. Eventually I gave up playing the violin. But it does not matter. Media matter does not matter. Occasionally I press the REC button on my recorder. With example four, *42° 7' N 19° 6' E Hour Eleven,* you hear the voice of my grandpa-in-law (Köner 2009). We were sitting on the veranda of his house in Montenegro, his wife had died and a bird in the pine tree seemed to answer in the pauses of his monologue, and I recorded that scene. This was in 2008. Grandpa is long gone now. I am holding the soft and warm hand of my radiant wife. Who is next? Life is a death camp. Maybe they are right, maybe it is a party. A rather small party, where everyone has left, and we stand alone, recognizing the void.

# References

Köner, Thomas (2004), *Banlieue du Vide*, Video Installation, quadrophonic sound. Available online: www.thomaskoner.com/bdv.

Köner, Thomas (2005), *Périphériques pt.3 Buenos Aires*, Video, 5min, stereo sound. Available online: www.thomaskoner.com/peripheriques3.

Köner, Thomas (2009), *42° 7′ N 19° 6′ E Hour Eleven*, Stereo sound, 5min. Available online: www.thomaskoner.com/montenegro.

Köner, Thomas (sound) and Jürgen Reble (film) (1992 ff), *Alchemie*, Performance. Available online: www.thomaskoner.com/alchemie.

# Index

# Concepts/keywords

Lightning Source UK Ltd.
Milton Keynes UK
UKHW020639201022
410784UK00012B/299